This book is to be returned or
the l

MATE

ADVA
SEN

Intelligent Processing of
MATERIALS
and
ADVANCED SENSORS

Proceedings of a symposium sponsored by the
Flow and Fracture and the Electrical, Magnetic
and Optical Committees, (ASM), held during the
Fall ASM-TMS meeting in Orlando, Florida, 1986.

Edited by:

H.N.G. Wadley
U.S. Department of Commerce
National Bureau of Standards
Gaithersburg, Maryland

Phillip A. Parrish
Defense Advanced Research Projects Agency
Defense Science Office
Arlington, Virginia

Bhakta B. Rath
Naval Research Laboratory
Washington, D.C.

Stanley M. Wolf
National Academy of Sciences
National Materials Advisory Board
Washington, D.C.

A Publication of The Metallurgical Society, Inc.

D
G70.28
INT

A Publication of The Metallurgical Society, Inc.
420 Commonwealth Drive
Warrendale, Pennsylvania 15086
(412) 776-9000

Printed in the United States of America.
Library of Congress Catalogue Number 87–42886
ISBN NUMBER 0–87339–070–9

Preface

Materials processing has, in recent years, become an increasingly important and challenging research field. It now is experiencing a breakthrough in control and automation that promises to play a central role in responding to rising concerns about United States international competitiveness in the materials marketplace. It is widely perceived that the recent poor performance of U.S. manufacturing stems, in part, from a failure to incorporate, at the rate of other countries, the latest developments in manufacturing science. If the U.S. industry is to restore its competitive edge, ways must be found to enhance materials quality and increase productivity while reducing manufacturing costs. The development and implementation (by industry) of better process control methodologies that exploit advances in process understanding, new process control sensors, and innovative process controllers, can potentially achieve these goals.

The Symposium on Intelligent Processing of Materials and Advanced Sensors was sponsored by the Flow and Fracture and the Electrical, Magnetic and Optical Committees of the Materials Science Division, (ASM), and held during the Fall ASM-TMS meeting in Orlando, Florida, 1986. The Proceedings consists of nineteen peer reviewed and edited papers presented during the Symposium. It covered a wide range of processes and encompassed metal, polymer and semiconductor materials. This Proceedings shows that for many important processes, the recent emergence of advanced sensors, predictive process models and new control scenarios (some featuring artificial intelligence technologies), has created the opportunity to significantly advance process control.

The Proceedings is divided into three sections. Section I is concerned with the development of advanced sensors for materials processing. Emphasis here was on emerging techniques for non-invasively probing the interior of bodies undergoing processing. Sensors for measuring internal temperature, instantaneous composition, polymer cure, alloy recrystallization etc. were discussed. Section II concentrated upon advances in process understanding (predictive modeling) where numerical simulation of materials processes is now able to provide a valuable tool for process optimization and a guide for optimum selection of sensors. It also promises to provide a quantitative basis for control decision making. Models for a broad range of processes were discussed--both batch and continuous casting, surface modification, sheet and plate rolling and seamless/welded tube formation. Section III is concerned with the integration of sensors, process models and new controllers capable of exploiting heuristic information. Examples of emerging approaches in both metal deformation and composite cure control are described.

In organizing this Symposium and in preparing this Proceedings we wish to acknowledge the central contribution of Antonette Nashwinter of NBS, who, together with Marlene Karl and Judy Parker of TMS, did in fact the majority of work associated with organizing the conference and publication of this Proceedings. We wish also to take this opportunity to express our gratitude to the authors, both for their clear and enlightened presentations, and their patience and timeliness in responding to the many queries and comments raised during the review and editing process.

Haydn N.G. Wadley
U.S. Department of Commerce
National Bureau of Standards
Gaithersburg, Maryland

Phillip A. Parrish
Defense Advanced Research Projects Agency
Defense Science Office
Arlington, Virginia

Bhakta B. Rath
Naval Research Laboratory
Washington, D.C.

Stanley M. Wolf
National Academy of Sciences
National Materials Advisory Board
Washington, D.C.

Table of Contents

PROCESS CONTROL

Advanced Sensors

NATURE AND CHARACTERISTICS OF SENSORS FOR INTELLIGENT

PROCESSING OF MATERIALS

Haydn N.G. Wadley

Metallurgy Division
National Bureau of Standards
Gaithersburg, Maryland 20899

Abstract

   To implement new process control strategies including Intelligent
Processing of Materials, advanced sensors are required to nonintrusively
evaluate process and microstructure variables. Researchers are
increasingly looking to innovative extensions of traditional
nondestructive evaluation technologies, such as ultrasonics and eddy
currents, for this. Examples of the nature and characteristics of
emerging sensors based upon these and other new measurement methods are
described. As these sensor development efforts evolve it is becoming
clear that their synergistic coupling with predictive process models
provides deep insights into the state of many processes. These
sensor/model combination approaches both offer opportunities to better
understand many important processes and to bring a hitherto unexpected
degree of control to materials processing.

3

Introduction

A sensor can be defined as a device that detects and measures a physical/chemical quantity and outputs it as an electrical signal. A commonly encountered example would be a piezoelectric sensor which detects mechanical stresses and outputs an electrical signal whose characteristics can be used to characterize the stress. Numerous sensors already exist, and are being increasingly used to measure quantities such as temperature, pressure, flow rate, pH etc. for an expanding multitude of purposes. For example, sensors are now extensively used in today's automobiles to control pollution, enhance fuel economy, implement anti-lock braking, activate safety measures during collisions, etc. Sensors have been used extensively in chemical engineering plants for process control; emerging devices promise to extend this use by measuring flow rates, particle distributions etc. Sensors are increasingly being incorporated in robotics for vision and tactile sensing where, when coupled with Artificial Intelligence techniques they enormously enhance the capability of robots.

The entire field of sensors is presently poised for major advances due to recent developments in non-invasive sensing technologies, increasingly powerful less expensive computing capabilities and emerging artificial intelligence techniques such as pattern recognition/expert systems. Technologies such as high intensity lasers, fiber optics, nonlinear optical materials, semiconductors, ultrasonics, eddy currents, etc. have rapidly progressed in recent years to the point where they now constitute a potential means to noninvasively probe the hitherto inaccessible changes and reactions of complex dynamically evolving systems. The rapid advance of expert system/pattern recognition software promise new control strategies capable of exploiting sensor outputs, predictive process models and computer accessible data bases for radical improvement of many existing manufacturing processes at a time of increasing concern about industrial competitivity in the international market place.

One of the most exciting areas of sensor development is that for the control of materials processing (1). Here, the juncture of emerging advanced sensors, recently developed predictive process models and expert systems has resulted in the possibility of an intelligent materials processing strategy. In this scenario, sensors continuously measure key process and microstructure variables. Predictive process models use these measurements to determine the current state of processing. Finally, control systems utilize the sensor outputs, predicted process conditions and the process models to adjust process variables to maintain the process within predetermined bounds and to feedback/forward through the process chain to accommodate for transient effects. It can readily be seen that the availability of inexpensive powerful computing and networking facilitates enables each step of a processing sequence to be integrated with the next and into a hierarchial factory automation scheme. Then every step could be optimally phased for maximum productivity and product flexibility. The possibility thus exists of applying to materials processing the principles of robotics and automation that are revolutionizing discrete parts manufacture and assembly. These new control strategies may even allow commercialization of new processes that produce advanced materials via process pathways hitherto considered too unstable for practical implementation.

This review assesses the unique nature of the advanced sensors needed to implement intelligent processing of **materials** (IPM) and other new control approaches. It describes the characteristics of those emerging from a few of the R&D efforts of recent years. It discusses the trade-offs between sensor needs and degree of process understanding and the synergy that may exist when both are combined for the control of materials processing.

<u>Sensor Needs for Materials Processing</u>

Determining the sensor needs of a process is not necessarily a trivial matter. The needs for sensors is linked to the level of process understanding (availability of predictive process models) and the degree to which the initial conditions of the process may vary (due to raw materials variability for example).

This can be exemplified by considering a model process, continuous metal casting, shown schematically in Fig. 1. For aluminum alloys, the direct chill casting process produces ingots of several meters at a time. In iron-base alloys, strand casters can operate continuously with the solidified product being fed continuously through a support roller system as shown in Fig. 1. In either process, the metal is poured from a ladle into a distributor or tundish from which it enters a water-cooled mold where surface solidification begins. By the time the metal leaves the mold a solidified "skin" has formed, typically of the order of several centimeters; sufficient to physically constrain the molten material within. Further direct water cooling removes heat through the "skin" and facilitates complete solidification.

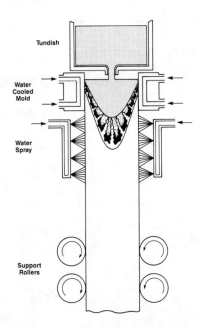

Figure 1 - Schematic diagram of continuous strand caster.

Numerous sensor needs exist for continuous casting, a partial list for the casting of steel is shown in Fig. 2. Included are sensors to assure melt cleanliness, determine oxygen content, liquid temperature, internal temperature distribution during and after solidification, etc. The performance required of these sensors depends upon the level of process understanding, the degree to which initial conditions can be maintained and the intended role of the sensor in the control loop; i.e., for direct process control or real time quality control.

5

Process Control Sensors

Sensors that measure quantities such as internal temperature distribution, composition, and microstructure are intended to be used to directly control the materials intrinsic properties in-process so that productivity and uniformity of properties is achieved. For example, chemical analysis during steel refining operations would be used to adjust composition to meet predetermined alloy specifications resulting in productivity benefits due to the avoidance of downgrading of heats and reduced processing times.

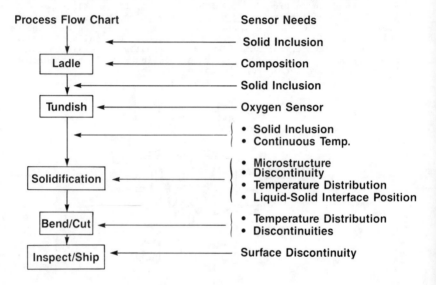

Figure 2 - Process flow chart and sensor needs for continuous casting of ferrous alloys.

In recent years, a great deal of progress has been made in process understanding and for many processes of import, sophisticated process models now exist. These models can be used to better identify key sensors and enhance the value of sensor data for process control purposes. For example, the contribution of process understanding to determining sensor needs for direct process control can be illustrated by considering internal tempering sensing. In the primary metals industry, an internal temperature distribution sensor, if available, would be the key sensor used to control the speed of a continuous caster. Presently, casters run at conservative speeds that are chosen on the basis of previous experience and solidification model predictions; they sacrifice casting speed (productivity) for the sake of eliminating the incidence of breakouts. Were the internal temperature mapped out at various cross sectional locations in the solidifying region, the *solidus* and *liquidus* isotherm positions and thus the thickness both of the solidified skin and "mushy" zone would be known together with the gradients (and thus effective heat removal rate) in the solid skin. This would enable optimum casting conditions to be maintained throughout a run. This would require, however, the

6

determination of temperature to ±5°C with at least 5 mm spatial resolution (30 x 30 pixels for a 150 x 150 mm cross section body); a considerable feat for any conceivable sensor.

However very little use of process understanding has been made in determining these needs. In fact, the cooling of a hot body is a very well understood process. The rate at which heat is removed is determined by a surface heat transfer coefficient, and the temperature gradients within are then dictated by well known physical laws of thermal diffusion. Even simple thermal models, incorporating an estimated heat transfer coefficient and temperature independent thermophysical constants, can predict the form of the isothermal contours within a solid body. Sophisticated process models embodying alloy solidification theory and heat flow have been developed for continuous casting. These, in principle, are capable of predicting all the required internal temperature distributions for control of the process. However, in practice, they are of only limited usefulness because the boundary (heat transfer) and initial (liquid temperature) conditions are sufficiently irreproducible and ill-defined that significant errors result in predicted isotherm location (and thus solid shell thickness) and they cannot, alone, be used for control purposes.

A hybrid approach utilizing a simpler sensor that measures in effect only the heat transfer rate and is coupled with predictive heat flow models may be the optimal solution to this process control need; especially if the initial temperature conditions are reasonably well defined, as they would be if the melt temperature were continuously monitored independently. Such a sensor could form the basis for casting speed control that would significantly enhance casting speed, and thus productivity, and provide a basis (from mushy zone shape) for controlling macrosegregation and related phenomena. Such a sensor approach is described later.

Quality Control Sensors

A second class of sensor needs are concerned with ensuring, on-line, quality rather than the direct control of the process itself. Advances in quality almost always lead to improvements in productivity and costs. Examples would be the real time detection of internal discontinuities (for example cracks, porosity and inclusions) as they form in alloys or dislocation generation during semiconductor single crystal growth. This can be seen because of the substantial savings that accrue if faulty conditions during processing are detected as they happen rather than at the completion of processing, or even after shipping, as is sometimes now the case. The benefits to a manufacturing economy of not having to discard or repair completed goods due to defects/out of specification primary materials would seem great enough to warrant a substantial sensor development effort to implement in-process quality control.

It is thus important to know the properties, location and formation mechanism of the defects of importance as a basis for identifying sensor needs. For example, thin strip metal destined for drawing/forming can suffer from the presence of subsurface inclusions. Strip containing these defects ruptures during subsequent forming operations due to inclusion nucleated ductile fracture. While inspecting material prior to shipment could avoid the problem, this wastes material and must drive up the cost of good strip. It might be much better to detect their occurrence earlier in-process ideally to stop their formation in the first place or at least reroute material prior to costly processing. The most obvious approach would be to detect inclusions as they form in the liquid prior to solidification. At this stage, steps still can be taken to stop their continued formation and for the removal of those already formed; for example, by using special mold fluxes during casting. The efficiency of these

7

fluxes is sometimes impaired, and it may be prudent to also monitor the
inclusion content in the just-solidified body. Thus, it seems necessary
to also develop sensors capable of detecting inclusions in hot solid
bodies very shortly after solidification. Process understanding is
obviously also important for this class of sensors, for it can again be
used to simplify the sensor by identifying the region where inclusions
are most likely to be deposited during solidification (quarter or center
line for example) and to thus limit the location of the region to be
probed by the sensor. Other quality control sensor needs include
internal and surface crack detection, porosity measurement, surface
hardness, dislocation content etc.

Emerging Sensor Methodologies

Without the emergence of better sensor systems, there can be no
real advancement in-process control, and without the advancements in-
process control, there can be no significant advancement in the
processes themselves. The development of process control sensors is now
becoming recognized as a critical factor in advancing materials
processing. The development of sensors is often complicated by a lack
of a broad enough knowledge base; the relationships between non-
invasive sensing mechanism/measurement and microstructure or defect are
yet to fully emerge. Furthermore, the hostile environment in which
sensors are used, the limited time available for measurements, the need
to avoid interference with the process itself, and the requirement of
acceptable cost/benefit ratios all introduce constraints upon practical
sensor systems that ultimately result in a less than ideal data set.
Thus, there exists a need to develop better models and algorithms of
sensor-material interaction, both for the analysis of limited data sets
and to refine the sensors themselves. The interplay between these
factors and its effect upon the sensor design can best be appreciated by
examining some emerging sensors.

SCHEMATIC OF SURFACE DEFECT SENSOR

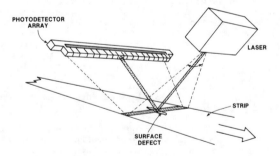

Figure 3 - Schematic illustration of surface defect sensor (after
reference 5).

Surface Defect Sensor for Strip Metal

The surface quality of steel and aluminum sheet is very important
to both the producers and users of these materials. Optical
reflectivity has been used successfully for surface inspection of slowly
moving sheet (3). Real time determination of surface quality prior to

8

coiling processed strip is extremely difficult because of the very high
strip speeds (up to two thousand meters per minute) and the wide variety
of imperfections that may occur. Furthermore, these imperfections must
be both distinguished from benign blemishes and characterized from one
another (4). In one development effort, the American Iron and Steel
Institute (AISI) is coordinating a collaborative program funded by a
consortium of steel and aluminum producers to develop a suitable sensor.

Research is being conducted by the Westinghouse Corporation and is
based upon a coherent light scattering approach (5), shown schematically
in Fig. 3. An intense collimated laser beam is rapidly scanned across
the width of the moving metal strip. Detector arrays are positioned
across the strip width at angles predetermined to optimize the defect
scattering (signal) to background (noise) ratio. The voltage outputs of
each detector are continuously digitized, digitally processed, and
pattern recognition and other techniques used to detect and characterize
defects. This information can be stored for each coil so that quality
is documented.

The combination of high strip speeds and many different types of
imperfection combine to pose major problems in high speed data
acquisition and digital signal analysis. For example, to fully inspect
a two meter wide strip moving at 30 meters per second (5500 feet per
minute) with a 1mm laser spot size requires state-of-the-art 8-bit
digitization rates of 200 MHz. Digital signal analysis of these very
dense data streams is too slow even with today's most advanced
computers, and preprocessing of the data is therefore essential. Only
then can pattern recognition and other AI software be used to
characterize the defects from the condensed data. Software alone cannot
solve this aspect of the problem however. Data preprocessing also may
be too slow or may filter out desirable signal traits. Devising sensor
measurement methodologies that simplify the needs for preprocessing and
defect characterization algorithms is a critical factor in the
successful development of this sensor, and indeed, many of the other
sensor needs of materials processing.

<u>Surface Modification Sensor</u>

Localized surface hardening through martensitic transformation of
iron base alloys is a frequently used method for enhancing resistance to
wear and fatigue (6). Directed high energy beams (laser on electron
beams) with energy densities of 10^3-$10^4 Wcm^{-2}$ are swept over the surface
causing transient local heating into the austenite phase field and
subsequent rapid cooling to ambient (7). This results in a surface
hardened layer, Fig. 4. The depth and hardness of the surface
modification need to be determined, preferably in-process, so that the
scan rate/beam shape can be adjusted to produce an optimal depth and
modified hardness.

Measurement of the velocity of ultrasonic surface waves as a
function of frequency is one method being explored to characterize
depth-varying properties (7). Surface (Rayleigh) waves have the special
feature that their amplitude decays rapidly with distance below the
surface on which they propagate. Furthermore, the rate of decay is
scaled by the wavelength, so that short wavelength waves decay rapidly
beneath the surface and therefore propagate at a velocity controlled by
near surface elastic properties. Long wavelength waves decay more
slowly with depth and thus propagate at a velocity controlled by a
combination of surface and substrate elastic properties.

Thus, as one increases the wavelength of a probe wave, a critical
wavelength is reached where the wave speed starts to change as the wave
begins to sample unmodified material beneath the depth of modification
(between 0.8 - 1.0 mm in Fig. 5). Furthermore, the difference in the
long and short wavelength velocity limits is a good indicator of the

hardness of the surface modified layer itself. The emergence of noncontact electromagnetic acoustic transducers (8) promises a means for non-intrusively making these measurements during processing. Better inverse algorithms for determining layer properties from wavelength velocity relations are, however, needed to improve the accuracy of the approach.

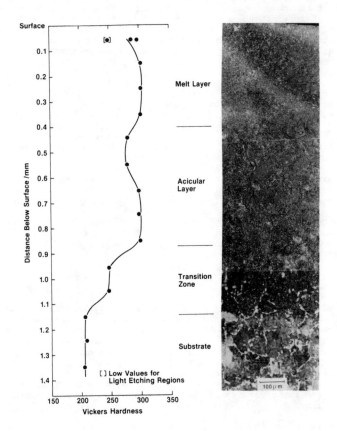

Figure 4 - Microstructure and hardness profile beneath an electron beam melted AISI 1053 steel surface (after reference 7).

Molten Metal Inclusion Sensor

The elastic constants and density of inclusions in molten alloys are different from those of the molten alloy itself. They thus have a different acoustic impedance. The difference in acoustic impedance of the inclusions results in scattering of incoming elastic energy waves, especially as their wavelength approaches the inclusion dimensions. This scattering can be measured and provides a convenient means for both detecting and gauging the number of inclusions present in a molten alloy and (by varying ultrasonic frequency) estimating their size.

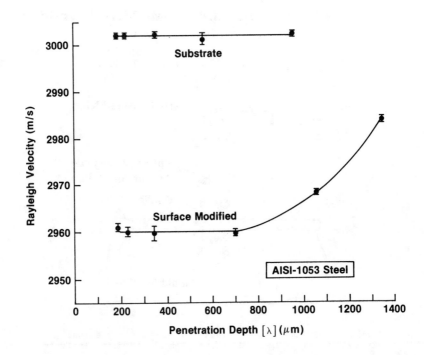

Figure 5 - Dependence of Rayleigh wave velocity upon wavelength (1) for a surface subjected to electron beam modification and for the unmodified surface (substrate) (after reference 7).

Mansfield et al. (9) have developed a sensor based upon this principle for aluminum alloys. The sensor consists of two flat parallel titanium plates that are fully emersed in the liquid metal. An ultrasonic pulse is applied to one plate and the ultrasonic reverberations between the plates monitored. When large numbers of inclusions are present, energy is scattered out of the forward propagating ultrasonic pulse, and the reverberations rapidly ring down. By measuring the attentuation at a fixed frequency (typically in the range 5-10 MHz), it has been possible to assess the melt cleanliness.

This ultrasonic approach is probably equally valid for steels and superalloys containing inclusions and/or ceramic particles. The higher temperatures involved pose a significant practical problem for ultrasonic generation/detection, but one that is probably amenable to solution especially given the availability of laser-generated ultrasound (10) and noncontact high temperature electromagnetic acoustic transducers (11).

11

Basic metals production could be significantly improved with the availability of a sensor capable of real time chemical analysis of molten metals and alloys (4). In an AISI collaborative program Kim et al. at Lehigh University are exploring the use of an *in situ* transient emission spectroscopy approach (12).

SCHEMATIC CHEMICAL COMPOSITION SENSOR

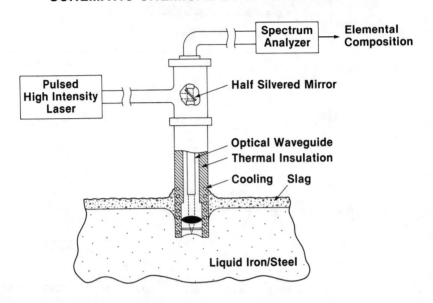

Figure 6 - Schematic illustration of the laser chemical composition probe being developed for liquid metal analysis.

In this approach, an intense laser pulse is used to evaporate and then electronically excite a small representative sample of the alloy within the melt, Fig. 6. The plasma thus created subsequently decays back to the ground state by the same photon emission processes traditionally utilized for emission spectroscopy analysis. The electromagnetic emission is collected and transported to a high-speed spectrum analyzer where the intensity of individual lines in the emission spectrum are used to infer the chemical composition with a millisecond time scale.

A key aspect of the approach concerns the degree to which the composition of the ablated material in the plasma truly is representative of the bulk composition. For instance, it obviously should not be heavily contaminated by slag. More insidious a problem is the potential for selective enrichment of the vapor by the more volatile elements present in the sample. The use of very brief, very high intensity laser pulses and careful calibration approaches show promise of overcoming this problem.

Internal Temperature Distribution Sensor

As discussed above, an internal temperature sensor is needed to image the temperature field within solidifying alloys so that solidification processing may be better controlled. One approach is to probe a body's interior with a penetrating radiation and attempt to measure a physical property that is temperature dependent. Ultrasound is one potential probe radiation, and ultrasonic velocity a measurable physical quantity that is strongly temperature dependent (13). For many metals the velocity is uniquely related to temperature and decreases with increasing temperature at a rate of between 0.5-1.0 $ms^{-1}°C^{-1}$, Fig 7. Velocity changes due to temperature gradients are usually much greater than those due to microstructure variations and internal stresses.

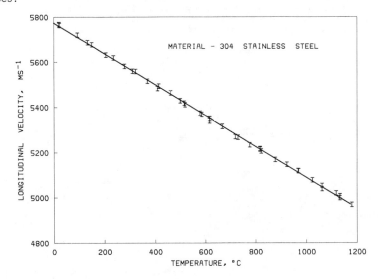

Figure 7 - Dependence of longitudinal ultrasonic velocity upon temperature for AISI 304 stainless steel (after reference 13).

At NBS, one approach to internal temperature sensing involves using an intense laser pulse to generate ultrasonic signals, and a high-temperature noncontact EMAT as a receiver, Fig. 8. The time-of-flight (TOF) of the ultrasonic pulses along paths of known length allows measurement of the average velocity along the path. Using reference data such as that shown in Fig. 7, this average velocity may be directly converted to average internal temperature.

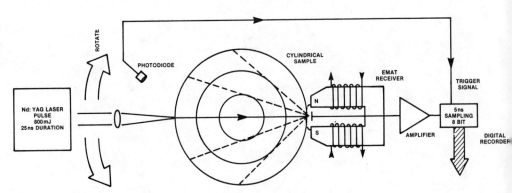

Figure 8 - Schematic diagram of ultrasonic internal temperature sensor utilizing laser generation/EMAT detection.

13

If independent TOF measurements for propagation along different ray paths are made, tomographic algorithms may be used to reconstruct an internal temperature image. However, it has been found better to use a least squares inversion procedure that facilitates incorporation of *a priori* information for the reconstruction. In this case, the *a priori* information is a thermal model which predicts the internal temperature distribution exactly provided the initial and boundary conditions are known. In practice these are not well defined, and so the TOF measurements are used for their determination. This is done by comparing measured with predicted TOF values, the latter based upon successive iterations of the temperature model with adjusted boundary conditions. Fig. 9 shows the good level of agreement between such an ultrasonic (curve) and embedded thermocouple (points) measurement of temperature for a 304 stainless steel sample.

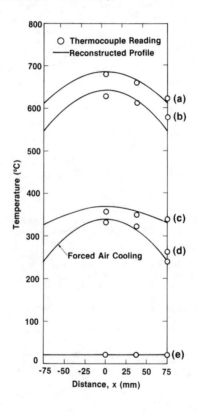

Figure 9 - Internal temperature distributions reconstructed from noncontact ultrasonic measurements compared with embedded thermocouple measurements (circles) for AISI 304 stainless steel. The distance corresponds to a line through the center of a 15cm x 15cm steel billet cross section (after reference 13).

Internal temperature sensors are needed for many processes such as high-speed aluminum extrusion, single crystal turbine blade growth, and the growth of single crystal semiconductors. Whilst the ultrasonic approach under development for steel may provide a sensor development route for these other processes, the different process constraints may dictate an alternative methodology. For example, the high local stresses associated with laser generation of ultrasound, whilst nonperturbing to the solidification of steel, may cause highly deleterious dislocation generation during semi-conductor single crystal growth. For aluminum extrusion, the ultrasonic velocity approach has poor precision due to the small part dimensions (a few millimeters thickness). The use of multifrequency eddy current techniques to measure electrical conductivity gradients is being explored as a sensor methodology for the aluminum extrusion sensor need (14). This approach could perhaps also be used to profile temperature during semiconductor single crystal growth since it induces very small stresses and semiconductors exhibit strong temperature-conductivity relations.

Summary

The emergence of advanced sensors coupled with process modeling and artificial intelligence/expert systems has created the possibility of new approaches to materials processing. Each of the processes controlled facilitate a fuller implementation of computer-integrated manufacturing where process, quality and product control and flexible manufacturing technologies may be merged into a single plant-wide system with enhanced productivity and product consistency while substantially reducing costs.

The sensor needs for materials processing are very demanding of today's measurement science. They are, however, intimately linked to the level of process understanding. Generally, the better the process is understood and capable of predictive modeling, the less stringent are the needs of sensors. Conversely, the more fully a process can be characterized by sensors, the less the dependence upon process models for process control. In devising the optimum control strategy for new processes it is important to assess the availability both of process models and sensors.

For materials processing, a premium is attached to those sensors that are capable of non-invasive probing of critical microstructure variables and quality factors that ultimately determine a product's performance. Ultrasound appears a particularly promising methodology because of its sensitivity to microstructure and the possibility of mapping internal temperature. New laser generation and electromagnetic acoustic detection schemes afford the opportunity for non-invasive sensing - a key factor for many sensor needs. Developments in eddy current, dielectric and microwave technologies are forming a basis for other equally useful sensor methodologies.

Materials processing is at the threshold of automation. If the potential of emerging sensors can be realized, the opportunities exist to revolutionize many materials processes and help reverse the declining competitivity of domestic materials by enhancing productivity and quality.

References

1. R. Mehrabian and H.N.G.Wadley, *Journal of Metals*, Feb. 1985, 51.

2. V.V. Horvath, NDE of Microstructure for Process Control, Ed. H.N.G. Wadley, *ASM*, 1985, 47.

3. L. Norton-Wayne, W. Hill and R. Brook, *Br. J. NDT*, *17*, 1977, 242-248

4. R. Mehrabian, R.L. Whiteley, E.C.van Reuth and H.N.G.Wadley, Eds."Process Control Sensors for the Steel Industry", (NBSIR 82-2618, U.S. Department of Commerce, 1982).

5. T. Porter, Private Communication.

6. B.H. Kear, E.M. Breinan and L.E.Greenwald, *Metals Technology*, Apr. 1979, 121.

7. B. Elkind, M. Rosen and H.N.G. Wadley, *Metall. Trans A*, 85, 261.

8. C.J. Morris and D.M. Keener, Proceedings Second National Seminar on NDE of Ferromagnetic Materials, Houston, 1986, 1.

9. T. L. Mansfield and C. L. Bradshow, NDE of Microstructure for Process Control, Ed. H.N.G. Wadley, *ASM*, 1985, 190.

10. G. Birnbaum and G.S. White, *Research Techniques in Nondestructive Testing*, Vol.VII, Ed. R.S. Sharpe, Academic Press, 1984, 259.

11. G.A. Alers and H.N.G. Wadley, *Progress in Quantitative Nondestructive Evaluation*, Vol. 6, Ed. D.O. Thompson and D. Chimenti, Plenum Press, 1987, In Press.

12. Y. Kim, This Proceeding.

13. H.N.G. Wadley, S.J. Norton, F.A. Mauer, and B.E. Droney, *Phil. Trans. Roy.Soc. Lond.* A 320, 341-361, 1986.

14 A.H. Kahn and H.N.G. Wadley, Proceedings of Aluminum Association Temperature Measurement Workshop, Atlanta, Georgia, May 1986.

MONITORING PIPE AND TUBE WALL

PROPERTIES DURING FABRICATION IN A STEEL MILL

G. A. Alers

Magnasonics, Inc.
215 Sierra Drive, SE
Albuquerque, New Mexico 87108

Abstract

Steel tubing has such a wide range of applications that it is produced almost continuously at speeds approaching 3 feet/second. Automatic process control and rapid inspection techniques must be used to keep production errors from generating long strings of faulty products and the inspection instruments must be designed to detect specific kinds of defects depending on how the tubes are produced. For example, seamless tubing formed by piercing a hot billet must have its wall thickness monitored over the full surface area of the pipe. Pipe formed by bending plate stock into a cylindrical shape and welding the seam must be inspected all along the joint to be sure that welding defects will not weaken the final product. Ultrasonic methods are the most promising of the possible NDE techniques for these inspections because they can interrogate both the I.D. and the O.D. as well as make accurate wall thickness measurements at production line speeds. Therefore, this paper will discuss recent developments in transducers and computer controlled instrumentation that are being tested in the steel mill environment.

Introduction

Steel tubes are used for piping in the oil, gas, and chemical industries as well as for structural members in the construction industry. As a result, the steel mills that fabricate all of this material must use high speed, continuous processing techniques that rely heavily on automatic control systems. This is particularly true for quality assurance procedures because modern applications are demanding tighter control over dimensional tolerances and rejection at ever smaller imperfection levels. In order to meet these requirements, new methods of inspection and the sensors that make them practical must be introduced and the computers that analyze the measurements must be speeded up and reprogrammed to display only the essential information to the operators.

This paper focuses on the inspection improvements that are being realized from the application of a special ultrasonic transducer that can operate at high speed on dirty, rough surfaces without precision alignment as well as withstand exposure to elevated temperatures. Thus, they are well suited to the steel mill environment. These advantages are being exploited to inspect the entire weld zone of ERW pipe, to detect and measure the depth of laps and seams on seamless tubing, to continuously monitor the wall thickness on heavy wall, large diameter pipes and to detect both transverse and longitudinal cracks on either the I.D. or the O.D. of the wall.

Fabrication Techniques for Tubular Products

There are basically two methods of forming cylindrical tubes out of the most common form of steel product generated in a high production rate steel mill. The most direct method is to pierce a hot billet with a long pointed tool to produce a hollow cylinder and then to work the cylinder to extend its length and reduce its wall thickness until the desired dimensions are reached. As a result, a pipe is produced that has no seam and very uniform metallurgical properties but which must be carefully inspected to expose variations in wall thickness because the tool cannot be held perfectly steady during the piercing operation. An often used alternate method of forming pipe is to roll a flat plate into a cylinder and weld the edges together to form a tube with a seam along its entire length. Such a product has very uniform wall thickness but it has a longitudinal seam with different metallurgical properties and the possibility of unacceptable weld defects. In either case, very critical applications such as deep oil wells, high pressure gas lines, and hazardous chemical conduits demand that no thin spots or internal defects be present in the pipe when it is installed. Thus, a very thorough inspection of each pipe being prepared for these critical structures must be performed either at the mill before shipment or at the construction site before installation.

At the present time, these detailed inspections are being performed using penetrating radiation from radioactive sources for wall thickness measurements and magnetic flux leakage or eddy current techniques for detecting laps, seams, and cracks on the I.D. and O.D. surfaces. Internal flaws such as laminations and lack of fusion, inclusions and porosity in the weld line are detected by ultrasonic techniques in which a water bath is needed to couple the transducer to the pipe. In all cases, the requirement for covering a large area with a small sensor has made the inspection process so slow that it cannot be performed on-line at the steel mill production rates. Thus, special inspection facilities are set up near the installation point and unacceptable pipes are returned to the mill for replacement. This adds considerably to the cost in time and money so there is an economic incentive for speeding up the inspection process enough to permit its performance at the steel mill at production line speeds of one to two hundred feet per minute (1.7 to 3.3 feet/second).

Ultrasonic Inspection

It has been known for some time that ultrasonic inspection techniques could be used to both measure the wall thickness and to detect flaws throughout the entire volume of the tube. To meet the inspection speed requirements, the pipe surface must be made wet by water squirters or by complete immersion so that piezoelectric transducers can couple the ultrasonic waves into and out of the steel. Furthermore, it takes a large number of transducers, each with a specific inspection task, arrayed around the pipe on precision holders to achieve adequate coverage of the surface area. Unfortunately, the mechanical alignment of all these transducers is quite critical so the installation and maintenance of a totally ultrasonic inspection unit is quite expensive even without considering the machinery to deliver the pipe to the transducer array and to hold it in position or even rotate it while the inspection takes place.

Recent developments in transducer technology have introduced an entirely new type of transducer that does not require the water couplant and is much easier to support mechanically because precise alignment with the surface is not critical. By using this transducer, the speed of inspection can be dramatically increased and the cost of the pipe handling equipment can be reduced considerably. Because the transducer operates by an electromagnetic induction process, large magnetic fields must be applied to the pipe and the signal strengths are small enough to require careful electronic design and installation procedures. Since the majority of steel tubular products are ferromagnetic, the necessary magnetic fields can be reached with very practical electromagnets and the current advances in solid state circuits, as well as the development of computerized data processing, make the electronic requirements easily attainable. The balance of this paper presents examples of how this new transducer is being applied to the production line inspection of tubular products in the steel mill.

Electromagnetic Transducers

The basic principles of the electromagnetic acoustic transducer or EMAT (1,2,3) are illustrated by the drawing shown in Figure 1. To form an ultrasonic transmitter, the wire is held close to the surface of a metal part and is driven with a large alternating current so that an eddy current will be induced to flow in the surface under the wire. When a large DC magnetic field is present, this eddy current exerts a mechanical force on the metal for the same reason that a current in the windings of an electric motor exerts a torque on the shaft of the motor. Since the drive current and the eddy current oscillate, the force also oscillates and launches an ultrasonic wave into the part. When this wave interacts with a flaw or encounters the opposite face of the part, a reflected ultrasonic wave returns to the region under the wire and causes the metal surface there to move. Just as in an electric generator, this motion in the presence of the DC magnetic field produces a surface current which in turn generates a voltage in the nearby wire by electromagnetic induction across the air gap. By amplifying this voltage, an echo signal can be presented on an oscilloscope just as ultrasonic echo signals are displayed in systems that use conventional piezoelectric transducers.

Therefore, an EMAT consists of a coil of wire held close to the metal to be tested plus a source of magnetic field while the usual transducer is formed by a piezoelectric ceramic disc coupled to the metal through a column of water. Although the latter device can produce larger ultrasonic signals, the former device can operate reliably over a part that is dirty or moving

19

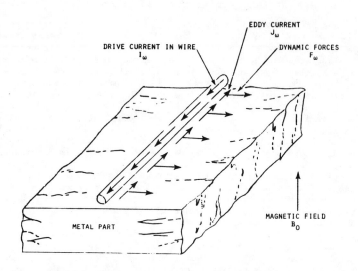

DRIVE CURRENT IN WIRE
I_ω

EDDY CURRENT
J_ω

DYNAMIC FORCES
F_ω

MAGNETIC FIELD
B_0

METAL PART

Figure 1. Schematic drawing of the essential features
of an electromagnetic acoustic transducer (EMAT).

rapidly past the transducer because the coupling medium does not have to be
carefully maintained. In addition, EMATs can excite a wider variety of sound
waves over a broader range of propagation directions (4). Thus, they can be
engineered to optimize their performance in any particular application.
Since the coil need only be near the surface and the sound originates from
the surface itself, precision mechanical supports for the coil are seldom
needed. The magnet, however, may be large and cumbersome but it need only
flood the area around the coil with a large magnetic field. Thus, it can
ride on wheels of its own or be attached to the floor and magnetize the pipe
through a large air gap. As the examples described below will show, the
addition of a few electromagnets to a steel mill full of massive pipe
handling equipment does not constitute a serious drawback.

Application to Seam Weld Inspection

    Safety codes often demand that the seam in welded pipe be inspected by
a pair of ultrasonic transducers that interrogate the joint from both sides
using shear waves that reflect into the weld region at an angle to the
surface. Figure 2 shows a schematic diagram of this inspection geometry with
a conventional transducer coupled to the surface through a wedge on the right
and an EMAT on the left. Two important advantages of the EMAT over the piezoelectric approach can be deduced immediately from an analysis of the
geometry of the angle beam technique. The first follows from the curved
surface of the pipe which bends the EMAT coil into a shape that focuses its
sound beam toward the I.D. while the refraction process at the wedge causes
the piezoelectric transducer's beam to be defocused into a diverging path
that could (in an extreme case) even cause part of the beam to fail to
reflect from the I.D. of the pipe. The second advantage in favor of the

20

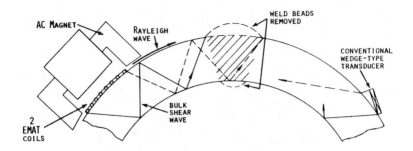

Figure 2. Schematic diagram of the preferred method for ultrasonic inspection of the weld region in seam welded pipe. The EMAT technique is shown on the left while the piezoelectric transducer method is shown on the right.

EMAT comes from the fact that a second coil can be placed in the magnetic field and a surface or Rayleigh wave can be launched toward the O.D. region of the weld without adding to the mechanical complexity of the transducer. This surface wave can be made very sensitive to O.D. flaws introduced by the removal of the weld bead or by faulty welding procedures. It also serves as a means of verifying that flaws detected by the angle beam inspection are located specifically on the I.D. or the O.D.

Figure 3 shows a photograph of an EMAT weld inspection system with the two transducers mounted on each side of the top of the pipe where the weld has been positioned by the pipe handling equipment of the steel mill. In this photograph, each EMAT has its own electromagnet that rides on the pipe on wheels located on the front and rear of the magnet housing. The EMAT coils themselves cannot be seen in this view because they are mounted under the magnets on a compliant support structure that bends the coils to the curvature of the pipe surface. Because of this flexible support, one EMAT coil can inspect a wide range of pipe diameters. Each magnet has two EMAT coils under it so that the weld is interrogated by two surface waves and two angle beam shear waves. The output of the data processing electronics is displayed on a two channel recorder driven by an odometer wheel that tracks the motion of the pipe past the inspection station. In this way, a flaw appears as a deflection of the chart recorder needle at a position on the chart that corresponds to the location of the flaw along the pipe. One channel records the largest signal seen by the two surface waves while the other channel records similar results from the two angle beam inspections. As usual, the amplitude of deflection of the pen is proportional to the maximum amplitude of the ultrasonic reflection from the flaw. An inspection system similar to the one shown in Figure 3 has been tested successfully in a steel mill in Ohio where the inspection speed exceeded 100 feet/minute.

Application to Longitudinal Flaw Detection

Because the EMAT coil conforms to the curvature of the pipe, the acoustic beam within the pipe wall is slightly focused and the size of the coil can be extended to cover more circumferential distance than a conventional piezo-electric transducer and wedge. As a result, each successive reflection of

21

Figure 3. Photograph showing two EMATs mounted to inspect a weld seam along the top of a pipe.

Figure 4. Photograph of an eight EMAT inspection system designed to give 100 percent coverage for I.D. and O.D. longitudinal defects on pipes ranging in diameter from 5" to 14" at speeds up to 60 feet/minute.

the sound between the I.D. and the O.D. can be made to overlap with its predecessor so that a flaw cannot escape detection by being off of the center line of the sound beam. This not only relaxes the requirement that the weld seam be maintained accurately at the midpoint between the two EMATs but it allows an individual transducer to detect flaws at a considerable distance around the circumference away from the transducer. By installing a modest number of EMATs and their magnets around the circumference, it is possible to inspect for flaws at any location around the pipe. Thus, an array of EMATs can provide 100 percent coverage for longitudinally oriented defects. Figure 4 shows a photograph of an eight EMAT inspection system designed to detect longitudinal flaws at any circumferential position on pipes in the diameter range from 5 inches to 14 inches. Note that each EMAT consists of an electromagnet on wheels that is the same as that used for the weld inspection shown in Figure 3. The electronic data processing equipment for this system (shown to the left of the circle of transducers in Figure 4), is capable of handling eight angle beam channels plus eight surface wave channels. The output data is displayed on a three channel strip chart recorder with one channel for I.D. flaws, a second channel for O.D. flaws and the third channel used to indicate the circumferential location of the flaw. If the flaw is detected by more than one EMAT, the circumferential location indicated is that associated with the largest signal.

Application to Lap and Seam Measurement

One of the most frequent kinds of flaws appearing in tubular products is the lap and the seam. They are insideous because they are crack-like defects that can penetrate deeply into the tube wall and be structurally very dangerous. Furthermore, their orientation can be such that an ultrasonic wave will be reflected away from the receiver and therefore missed in an ultrasonic inspection. Furthermore, the opening at the surface can be so small that there is very little flux leakage and the flaw can go undetected during a magnetic inspection.

Since surface or Rayleigh types of ultrasonic waves cling to the surface and penetrate below it to a depth comparable to their wavelength (0.02 to 0.2 inches, i.e., 0.6 to 6 MHz), they are very well suited to the interrogation of flaws that lie within this surface layer. Furthermore, EMATs are able to excite and detect these waves much more reliably than a piezoelectric wedge-type of transducer because the latter device must be coupled to the surface through a liquid or grease layer that cannot be constant on irregular surfaces. Thus, EMAT ultrasonic inspection systems designed to use surface waves such as those shown in Figure 3 and 4 are very well suited to the detection of laps and seams. When a surface ultrasonic wave interacts with a lap or seam, either the reflected or the transmitted wave can be used as an indicator of the presence of the flaw (5). The reflected wave is not very reliable for quantitative measurements because its amplitude as observed by a fixed receiver is very sensitive to the angular orientation of the lap or seam. The amplitude of the transmitted wave, on the other hand, depends primarily on how far the crack tip lies below the surface and not directly upon the angular orientation of the crack. This useful result can be rationalized by viewing the transmitted wave amplitude as a quantitative measure of how much of the ultrasonic beam is intercepted by the crack, i.e., how extensive a shadow is cast by the lap or seam. Theoretical analysis (5) of this shadow casting interaction between surface waves and inclined cracks is available and can be used for quantitative determinations of the dimensions of laps and seams from measurements of the change in transmitted signal amplitude when a lap or seam is present between a Rayleigh wave transmitter and receiver. This method for both detecting and measuring laps and seams was evaluated in an extensive study of the various modern methods for inspecting bars and tubes at production line speeds (6). By performing

23

destructive tests on the samples that showed flaws, it was possible to make a comparison between the flaw depth predicted by the various inspection methods and the true flaw depths. Thus, the quantitative capabilities of each NDE technique could be tested in a very accurate way. Figure 5 shows the final results of this extensive research program and it is clear that the technique based on EMAT surface waves was outstanding in its ability to predict the true dimensions of the flaws over a wide range of flaw sizes. As a result of these tests, a prototype EMAT inspection system is being used for the high speed inspection of bars in a pilot plant located in central Ohio. This prototype unit is also noteworthy for its ability to cover the entire circumference of the bars without rotating either the sensors or the bar during inspection.

Measurement of Wall Thickness

As pointed out in the introduction to this paper, the most serious inspection problem facing the manufacturer of seamless tubing products is the monitoring of the wall thickness. Because of the method used to fabricate this kind of pipe, the wall can have an eccentricity of the I.D. which causes one side of the pipe to be thin and the opposite side to be thick. It can also have localized thin and thick spots. Safety codes usually demand that the combination of these two "defects" shall not cause any point on the pipe to have a wall thinner than 87-1/2 percent of the nominal thickness. As a result, it is important to make an accurate measure of the thickness of the pipe wall at as many points on the pipe as is practical in the time available. For large diameter pipes being produced at speeds approaching 200 feet/ minute, the monitoring of wall profiles is a difficult problem that demands very rapid inspection techniques and many sensors around the circumference. Only an EMAT-based thickness gaging system has been able to approach the coverage and speed requirements for an on-line installation inside a steel

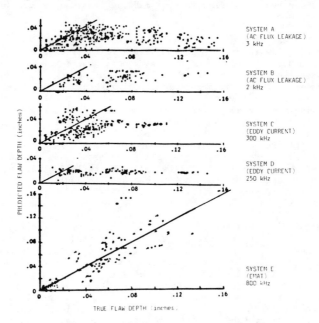

Figure 5. Comparison between actual flaw depth and the flaw depth indicated by various NDE techniques. (Results taken from Ref. 6.)

mill. Such a system has been installed at a mill in northern Ohio. It
consists of an array of eight retractable electromagnets arranged as shown
in Figure 4, around a circle capable of accommodating tubes in the 10 to 26
inch diameter range. Each magnet was equipped with a special EMAT coil that
would allow the transit time of an ultrasonic pulse through the pipe wall
under the transducer to be measured. Since such a measurement was performed
every 1/8 of an inch along 8 paths down the length of a 40-foot long pipe, a
dedicated computer was installed to process the transit time data and record
final thickness values in the computer memory as well as signal the operator
whenever too thin a wall was discovered. Once all the data was stored, the
printer could interrogate the memory and generate a hard copy presentation
of the thickness profiles observed on any particular pipe as well as any
other dimensional parameters that could be deduced from the thickness measurements. Figure 6 shows an example of one of these printouts wherein the
operator has chosen to see a graph of the minimum wall thickness, the
eccentricity, the average wall thickness at each circumference and examples
of two specific wall profiles observed by two of the eight EMAT sensors. In
addition to this delayed report, the computer made a real time printout of a
table of specific dimensional characteristics on each joint of pipe inspected
so that the mill management and the operator could quickly respond to trends
in the production run of a large number of pipes.

Application to Transverse Flaw Detection

Although the manufacturing process tends to elongate flaws parallel to
the length of a pipe to produce a predominance of longitudinally oriented
defects, it is also possible to develop cracks and pits that run in a transverse direction around the circumference of a pipe. Thus, a complete inspection system requires an array of sensors oriented to detect this type of flaw
also. Ultrasonically, this means that the acoustic wave must propagate along
the length of the tube and 100 percent coverage can be obtained only if a
large number of EMATs are arrayed around the circumference or if a few EMATs

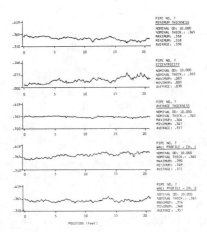

Figure 6. Example of the data displays possible from a multi-
channel EMAT thickness gage installed in a seamless tube mill.

are mounted on a rotating machine to scan the sensors around the pipe as the pipe moves past the inspection station. This latter alternative can be very expensive because of the mass and speed of the rotating parts. However, the former approach only requires repetitive production of identical electronic channels. The simplicity of the fixed array of EMATs concept can be seen in Figure 7 which shows one EMAT sensor riding on the pipe inside a single electromagnet that encircles the tube. Because the pipe is ferromagnetic, most of the magnetic flux produced by this coil is concentrated in the pipe wall and the mechanical support structure that is attached to each EMAT sensor can utilize as much of the ample space between the magnet and the pipe as is required. Not only can the sensor shoe that rides the pipe contain transverse flaw detection EMAT coils but it can also hold coils that excite surface waves and wall thickness measuring EMATs. A 32 channel thickness gaging system designed to give 100 percent coverage of pipes in the 7 to 10 inch diameter range is being tested at a pipe storage and repair facility in Louisiana. Its output demands a dedicated computer and large disc storage capability to handle all of the data produced on one joint of pipe. At a throughput speed of 40 to 60 feet per minute, the set of 32 individual thickness contour plots must be generated by a printer operating in an off-line mode in order to yield a hard copy display of the wall thickness map of the entire pipe. An example of this 32 channel map is shown in Figure 8.

Conclusions

In order to respond to worldwide competition in the production of steel, the United States steel industry has been driven to introduce more and more automated processing methods. This has, in turn, caused a demand for new sensors and computerized data processing techniques that can operate under the conditions that exist in a steel mill. This paper has reviewed the state-of-the-art in the manufacture of steel pipes and tubes and has demonstrated by example how the introduction of a new ultrasonic transducer has enabled on-line quality assurance procedures to keep up with increased production speeds.

References

1. R. Bruce Thompson, "Noncontact Transducers," Proc. 1977 Ultrasonics Symposium, IEEE Cat. No. 77 CH1264-ISU, pp. 74-83 (1977).

2. G. A. Alers, "Applications of Electromagnetic Acoustic Transducers (EMATs)," Proceedings of 26th National SAMPE Symposium, vol. 26, pp. 34-44 (1981). Society for the Advancement of Materials and Process Engineering, PO Box 613, Azusa, CA 91702.

3. B. W. Maxfield and C. M. Fortunko, "The Design and Use of EMATs," Materials Evaluation, vol. 41, pp. 1399-1408 (1983).

4. C. F. Vasile and R. B. Thompson, "Periodic Magnet Noncontact EMAT—Theory and Application," Proc. 1977 Ultrasonic Symposium, IEEE Cat. No. 77 CH1264-ISU, pp. 84-88 (1977).

5. B. Q. Vu and V. K. Kinra, "Diffraction of Rayleigh Waves in a Half Space. II. Inclined Edge Crack," J. Acous. Soc. Am., vol. 79, pp 1688-1692, June 1986.

6. R. Palanisamy, et al, "On the Accuracy of AC Flux Leakage, Eddy Current, EMAT, and Ultrasonic Methods of Measuring Surface Connecting Flaws in Seamless Steel Tubing," Review of Progress in Quantitative NDE, vol. 5A, pp. 215-223, D. O. Thompson and D. E. Chementi, Editors, Plenum Press, New York, 1986.

Figure 7. Photograph of an encircling electromagnet with one EMAT suitable for detection of transverse flaws along a line down the total length of tube.

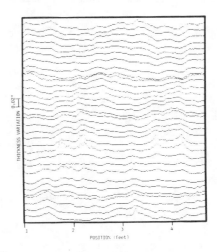

Figure 8. Example of a computer printout from a prototype 32 channel EMAT thickness gage designed for 100 percent coverage at high throughput speeds.

ULTRASONIC SENSORS FOR CHARACTERIZATION OF PHASE

TRANSFORMATIONS

M. Rosen
Materials Science Department
and Center for Nondestructive Evaluation
The Johns Hopkins University
Baltimore, Maryland 21218, USA

Abstract

Application of contact (piezoelectric) and noncontact (laser generation and detection) ultrasonic techniques for dynamic investigation of precipitation hardening processes in aluminum alloys, as well as crystallization and phase transformation in rapidly solidified amorphous and microcrystalline alloys will be discussed. From the variations of the sound velocity and attenuation the precipitation mechanism and kinetics were determined. In addition, a correlation was established between the observed changes in the velocity and attenuation and the mechanical properties of age-hardenable aluminum alloys. The behavior of the elastic moduli, determined ultrasonically, were found to be sensitive to relaxation, crystallization and phase decomposition phenomena in rapidly solidified metallic glasses. Analytical ultrasonics enables determination of the activation energies and growth parameters of the reactions. Therefrom theoretical models can be constructed to explain the changes in mechanical and physical properties upon heat treatment of glassy alloys. The composition dependence of the elastic moduli in amorphous Cu-Zr alloys was found to be related to the glass transition temperature, and consequently to the glass forming ability of these alloys. In this review application of Rayleigh surface waves is shown to enable nondestructive determination of the elastic properties and thickness of electron-beam surface modified layers. Dynamic ultrasonic analysis was found to be feasible for on-line, real-time, monitoring of metallurgical processes.

Introduction

Ultrasonic nondestructive evaluation has traditionally been concerned with the search and location of flaws in materials structures and the determination of their distribution and orientation. A considerable body of knowledge has also been accumulated from ultrasonic scattering studies in assessing grain size and orientation effects in materials. Recently, it has become widely recognized that ultrasonic measurements can be used to characterize materials structures and hence properties such as strength, toughness, effect of residual stresses so as to supplement, or even replace, the conventional destructive techniques employed in metallurgy. Ultrasonic nondestructive characterization offers distinct advantages in that materials properties can be verified on actual components of engineering structures.

29

The scientific literature is extremely scarce in dynamic nondestructive characterization (NDC) whereby ultrasonic techniques are applied to on-line, real-time, monitoring of microstructures for the control of metallurgical processes. Inherent difficulties related to these types of measurement methodologies include operation at elevated temperatures and in hostile environments. Furthermore, there are little basic data available on the relationship between the metallurgical microstructures and measured ultrasonic responses. Therefore, the full potential of ultrasonic techniques in this field is yet to be realized.

Many advantages single out the ultrasonic approach to sensors compared with other techniques. The following attributes of ultrasonic sensors are the most prominent:

- Simultaneous acquisition of two types of data: a) Sound wave velocity, related to the elastic properties of the material and b) ultrasonic attenuation, containing important microstructural information connected with the mechanical, physical and chemical properties of materials and processes.

- High sensitivity to occurrence of phase transformation, microstructural and morphological changes.

- Nonintrusive, when ultrasonic energy of low strain amplitude is applied. However, by using a high strain amplitude (at appropriate ultrasonic wave frequency) the threshold can be determined at which the sound wave becomes an active participant in the metallurgical process.

- Noncontact and noninvasive, when high power pulsed lasers are used for thermoelastic generation of ultrasound and laser-interferometric techniques are employed for the detection of the propagating waves. By means of this methodology the ultrasonic sensor approach can be applied to materials and processes located in hostile environments e.g., high temperatures and pressures, corrosive atmospheres, etc.

- Remote operation, applying lasers for generation and detection in conjunction with bundles of optical fibers. In this case, the lasers can be located at a safe distance from the hostile environments. The fiber bundles transport the ultrasonic energy to and from the material subjected to thermomechanical treatment.
- The acquired data (sound wave velocity and ultrasonic attenuation as a function of different variables) can be utilized as sensors for artificial-intelligence processing of materials since the data can be:

  . digitized
  . converted to electrical impulses
  . analyzed according to a predetermined procedure
  . rerouted for monitoring and control

Recent studies have addressed the specific issue of dynamic, real-time nondestructive characterization of metallurgical processes, e.g., precipitation hardening of aluminum alloys, crystallization of amorphous alloys and determination of properties on surface-modified materials (1-4). For the determination of the kinetics and mechanism of precipitation hardening and their relationship with observed mechanical properties, conventional bulk ultrasonic techniques were adapted to continuously monitor changes (1). However, in the case of metallic glasses and other amorphous structures where the samples are in the form of thin ribbons or thin layers on thick substrates, alternative methods had to be developed and applied. One of the techniques employs a coil to magnetostrictively launch and detect extensional waves in a ribbon or magnetic material, called the "driver". The specimen can be coupled to the end of the driver, producing an echo pattern from which the extensional velocity in the material can be derived by a pulse-superposition technique(5). A pulser-receiver was constructed to feed high current pulses to the coil and to amplify the electrical signals produced in the coil by acoustic waves. However, in attempting to dynamically characterize metallurgical processes in amorphous ribbons or structurally modified thin surface layers, these ultrasonic techniques were found to be inadequate. Thus, a new approach for contactless generation and detection of acoustic waves, using a laser generation and laser interferometric detection system was developed (6).

Ultrasonic Waves: Generation and Detection Techniques

The relationship between experimentally determined sound wave velocities (longitudinal and transverse) and the elastic moduli, is expressed in terms of the equations of motion of elastic waves in a solid, and Hooke's law for an isotropic medium. Thermodynamic formalism shows that a solid should exhibit an anomalous elastic behavior ("softening effect") in the vicinity of a phase transformation point. Anomalous changes in the elastic moduli are expected to occur, to a different degree and magnitude, in the proximity of metallurgical reactions involving precipitation processes, segregation of solute, magnetoelastic effects accompanying magnetic ordering and other phenomena.

The propagating sound waves in the solid undergo a series of loss processes due to scattering by reflection and refraction from grain and phase boundaries, thermoelastic losses including elastic anisotropy effects in polycrystals (Zener effect), as well as attenuation of the sound wave energy due to interaction of the propagating wave with electrons and phonons. Interaction with dislocations in the crystal lattice is also responsible for the observed sound-wave damping effects. In instances where dislocation movement plays a prominent role, e.g., in martensitic phase transformations, the behavior of the ultrasonic attenuation may contain important information with respect to the transformation mechanism. During precipitation hardening of 2024 aluminum alloy the main contributions to observed changes in attenuation (1) were found to be due to geometric scattering effects by newly formed precipitates of a certain critical size distribution, as well as to resonant interaction between the propagating sound waves and the dislocation loops generated around precipitates.

For bulk ultrasonics, where the wavelength of the propagating waves is much smaller than the specimen dimensions, the velocity and attenuation of ultrasound in specimens can be determined through the use of pulse-echo-echo overlap technique (5). The ultrasonic transit time, between the opposite faces of a flat and parallel sample, is determined by overlapping two successive members of an echo train on an oscilloscope display. This method, by stroboscopic identification, allows the time interval between pulses to be determined to within 1 part in 10^4. The logarithmic decay of the amplitude of successive pulses determines the ultrasonic attenuation. The apparatus for measuring both velocity and attenuation is shown in Figure 1.

Dynamic measurements of sound velocity and ultrasonic attenuation, during metallurgical changes in the temperature interval between the ambient and $250^{\circ}C$ were performed while the specimen and the ultrasonic sensors were in direct contact in an oil bath. For static ultrasonic measurements, carried out at room temperature, with the purpose of correlating ultrasonic data with mechanical properties, both the samples and the sensors were immersed in a distilled-water tank at a constant separation. Careful adjustments were made to ascertain an exponential decay of the amplitude of successive echoes, combined with a maximal number of echoes in the pulse train.

The growing interest in metallic glasses has created the need for the determination of their physical and mechanical properties. Some of these properties, especially the mechanical ones, cannot be measured conveniently by traditional ultrasonic methods as the high cooling rates required to form metallic glasses restrict their physical shapes to shallow layers or thin ribbons. Alternate measurement methodology needed to be developed in order to accurately assess these properties. One of the techniques employs a coil to magnetostrictively launch and detect extensional waves in a ribbon magnetic material, called the "driver." The specimen can be coupled to the end of the driver, producing an echo pattern from which the extensional velocity in the material can be derived by a pulse-superposition technique(5). A pulser-receiver was constructed to feed high current pulses to the coil and to amplify the electrical signals produced in the coil by the acoustic waves.

The noncontact feature of both generation and detection of ultrasonic waves can be very advantageous in situations requiring physical separation between measuring system and the material under investigation, for instance, when high temperatures or hostile atmospheres are involved. Furthermore, the contactless generation and detection precludes interaction with, and modification of, the wave propagation pattern under the study. In addition, laser generation of acoustic waves yields a wide variety of propagation modes (longitudinal, transverse and plate modes, Rayleigh waves) over a wide frequency range, thus enhancing the amount of information obtained from a single measurement and rendering the sensor more compatible to a wide range of experimental situations. Compressive stress waves that propagate in a material can be generated by transient loads applied by rapid energy transfer from single-pulse Q-switched high-energy Nd:YAG laser, (Fig. 2). Propagation of ultrasonic waves in a medium causes surface displacements on the material that can be measured optically by exploiting the phase shift of an optical beam reflected from the surface of the material. When the reflected beam is combined with a reference optical beam, from a helium-neon laser, optical phase changes are converted into amplitude phase changes that are detectable by a sensitive photodiode. These variations in amplitude are proportional to the surface displacements on the specimen. Potential problems arising from the fact that phase changes also result from relative motion among optical components of the system and from temperature and pressure fluctuations of the ambient air, are

VELOCITY

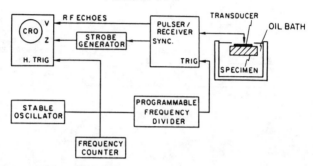

(a) THE VELOCITY-MEASURING SYSTEM (CRO, CATHODE RAY
OSCILLOSCOPE).

ATTENUATION

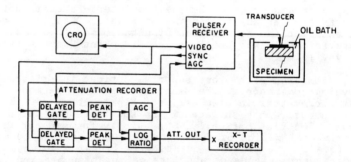

(b) THE ULTRASONIC-ATTENUATION-MONITORING SYSTEM.

**THE FACILITY FOR INSITU MEASUREMENT OF
LONGITUDINAL OR SHEAR VELOCITY AND ATTENUATION.**

Figure 1

33

SOUND VELOCITY MEASUREMENT SYSTEM

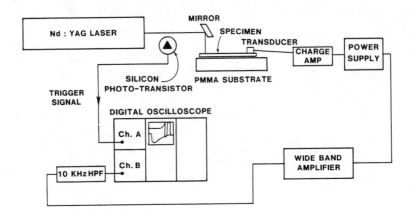

Figure 2.

prevented by appropriate design of the interferometer. Optical
schemes due to Fizeau and Michelson are particularly suitable
for the specific purposes. In the Fizeau version, the optical
probes are separated to allow accurate measurements of travel
time of an ultrasonic wave in the materials over a well-defined
distance. Furthermore, the variation in magnitude of the surface
displacements detected by the two interferometers determines the
ultrasonic attenuation in the material. Thus, both velocity and
attenuation can be measured simultaneously (Fig. 3). The Fizeau
interferometer was built at Johns Hopkins University by Dr.
Harvey Palmer.

Compared with other sensors, optical interferometers offer
several advantages. The sensitive area can be made very small, a
few micrometers in diameter, for highly localized measurements.
Consequently, the surface of the specimen does not necessarily
have to be flat optically. The highly focused optical beams
permit utilization of specimens with conventionally machined
surfaces. The measured quantity is linear displacement.
Independent methods of absolute calibration are applicable.
Bandwidth, determining the fidelity of reproduction of signal
waveforms, is not limited by the character of the transduction
process but by the electronics of the interferometer detectors.
Therefore, performance can largely exceed that of conventional
piezoelectric transducers. Small signal resolution and bandwidth
are related, thus linear displacements of a few angstroms are
detectable at 7 MHz bandwidth.

Laser pulse irradiation produces a stress pulse of short
duration (15 ns) and relatively high amplitude (up to 200 mJ
power) making the investigation of very thin (about 20 μm) and
highly attenuating specimens possible. Since a multitude of
acoustic wave propagation modes is generated, the dual

34

Overall Optical System

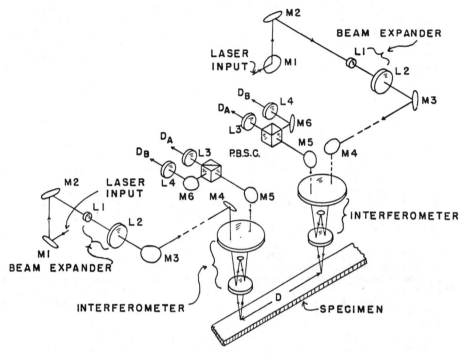

Figure 3   DUAL LASER INTERFEROMETER.

interferometer provides important information concerning the elastic and anelastic properties of the material under static or dynamic conditions.

The contactless generation and detection of acoustic waves is extremely advantageous because specimens can be studied while they are subjected to thermomechanical processing under adverse environmental conditions. No transducer protection (e.g., from elevated temperatures) is necessary, and the detected ultrasonic waveform is unperturbed by extraneous effects due to physical contact between specimen and transducer. Data acquisition is straightforward and real-time or post-test analysis is feasible. The capability to obtain the frequency dependence of both sound velocity (in dispersive regimes) and ultrasonic attenuation offers new opportunities in solid state and physical metallurgy research where the elastic properties and acoustic energy absorption play a prominent role in the characterization of the processes.

Laser and electron beam irradiation techniques are being extensively applied for the modification of surface properties of metallic structures (7,8). The nature of the modified surface zones is not amenable to conventional nondestructive characterization. However, the analysis of Rayleigh wave velocities, and the determination of the elastic moduli, may lead to a better understanding of the properties of the modified

surface layers. Laser or piezoelectrically generated Rayleigh
surface waves probe preferentially the near subsurface region of
the sample, and are extremely sensitive to variations in the
elastic properties and ultrasonic attenuation of the material
medium. The extent of penetration of the Rayleigh surface wave
normal to the surface of the sample depends on the frequency of
the propagating Rayleigh waves and the exponential decay of their
intensity. For a typical
metallic phase, the Rayleigh wave velocity, V_R is about
3,000 ms^{-1}. For a frequency of 10 MHz, the Rayleigh
wavelength is, therefore, 300 µm. Theory shows that 90 percent
of the energy of the Rayleigh wave is contained within one
wavelength from the surface. Thus, the sub-surface region can be
monitored with a high degree of accuracy. Moreover, Fourier
analysis of the frequency content of the Rayleigh waves, when the
dual-laser interferometer is used to contactlessly detect the
propagating surface waves, allows the precise determination of
the thickness of the surface layer. This gauging procedure is
possible because of the unique property of the Rayleigh wave
velocity that is independent of frequency.

Two properties of the Rayleigh surface wave make its
detection possible by optical means. One is the surface
microcorrugation (misrodistortions) as the Rayleigh wave
propagates through the material medium. Second, is the periodic
variation of the index of refraction while the wave propagates on
the surface. By means of optical interferometry, e.g., dual
Fizeau interferometer, these waves can easily be detected in a
contactless fashion. Thus, the contactless generation and
detection of Rayleigh waves enable one to dynamically study the
near-surface modifications of the material while the samples are
contained in a hostile environment, and are subjected to
programmed thermomechanical treatments subsequent to laser or
electron beam alloying and microstructural modification (Fig. 4).

Nondestructive Characterization of Precipitation Hardening Phenomena in Aluminum Alloys

Precipitation hardening, or aging, is an important
metallurgical process whereby the material's strength and
hardness can be augmented, in several instances, by an order of
magnitude. Precipitation hardening is a thermally-activated,
diffusion-controlled, process of particular importance in copper,
iron, and aluminum-base alloys. The improved mechanical and
physical properties of these alloys depend, largely, on the
microstructure, spacing, size, shape, and distribution of the
precipitated particles, as well as on the degree of structural
and crystallographic coherency of the particles with the matrix.
For example, the precipitation hardening process in aluminum
alloys involved formation of Guinier-Preston (G.P.) zones, and
particles that are generally coherent of semicoherent with the
matrix and contribute to the strengthening of the alloy. At
higher temperatures incoherent particles form, that dramatically
reduce the strength and hardening of the alloy. Of particular
importance in aluminum-base alloys are the θ' particles,
initially semicoherent with the matrix, that grow in size as the
precipitation process progresses. At a certain critical size,
when the stress fields surrounding these precipitates are

36

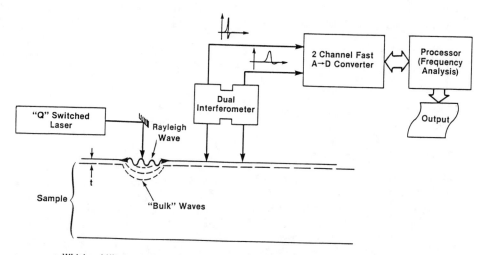

- Wideband Ultrasonic Pulse Generated by Laser Pulse.

- Vertical Displacement v.s. Time is Measured at Two Different Points by Two Wide Band Optical Detectors (Interferometers).

- Frequency Analysis of the Two Pulses Gives Attenuation v.s. Frequency and Velocity v.s. Frequency Curves.

- Since the Penetration of Rayleigh Waves is Frequency Dependent, the Thickness t of the Amorphous Layer can be Evaluated From These Curves.

RAYLEIGH WAVE VELOCITY MEASUREMENTS IN MICROSTRUCTURALLY
MODIFIED LAYERS EMPLOYING LASER GENERATION AND LASER
INTERFEROMETRIC DETECTION OF ULTRASONIC WAVES

Figure 4

maximal, dislocation rings (Orowan loops) form around them causing the coherency to be weakened. At this point overaging, or softening, begins.

Of great technological importance is the development of a nondestructive sensor that will permit: (a) determination of the extent of matrix supersaturation in alloys prepared by new processes such as rapid solidification, and (b) real-time monitoring of the precipitation hardening process to determine particular events such as the instant when particles lose coherency with the matrix and overaging begins. Of course, solution of the latter problem implicitly addresses the former.

Figure 5 shows the time dependence of the changes in the sound velocity of 2219 aluminum alloy during isothermal aging at three temperatures. The nondestructive characterization of the aging process, as manifested in Figure 5 and corroborated with optical and electron microscopy, hardness, strength, and eddy currents conductivity data, shows a series of dips that correspond to the maximal rate of formation of θ'' and θ' particles.

37

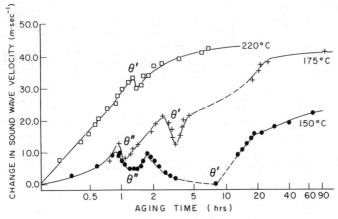

ULTRASONIC VELOCITY DURING THREE ISOTHERMAL HEAT
TREATMENTS OF 2219 ALUMINUM ALLOY

Figure 5.

The prominent peaks in ultrasonic attenuation, Figure 6,
correspond to the peaks in hardness and represent the loss of
coherency of particles with the matrix and beginning of the
softening process.   The peaks in ultrasonic attenuation arise
from an interaction between the acoustic vibrations of the
ultrasonic waves propagating through the material and the
interface dislocations surrounding the   particles.   The strength
of the interaction will depend on the total volume of the
particles, their size, shape and distribution.   The peak value of
attenuation can be qualitatively related with the total volume of
coherent precipitates that lost coherency with the matrix.   The
temperature dependence, indicated by the shift in the attenuation
peaks (Figure 6) permits the determination of an activation
energy that is found to be very close to that obtained from the
shift of the hardness peaks (1).   Corroborative evidence obtained
by means of electron microscopy and hardness substantiates the
proposed mechanism of coherency loss in this alloy.

This study demonstrates the feasibility of the ultrasonic
NDC technique to determine the time of formation of the various
precipitates, the kinetic parameters of the process, as well as
to establish, by means of ultrasonic attenuation, the time when
the alloy begins to suffer loss of its mechanical properties due
to overaging.

Nondestructive characterization measurements have been
carried out on 160 Al-Cu alloy samples subjected to a series of
different preaging heat treatments prior to thermal processing to
different tempers.   For each temper the maximum hardness of the
alloy was found to correlate to a particular value of the sound
wave velocity.   In addition, ultrasonic attenuation was found to
consistently decrease in a linear fashion as hardness is
increased, Figures 7 and 8.   This investigation demonstrates the

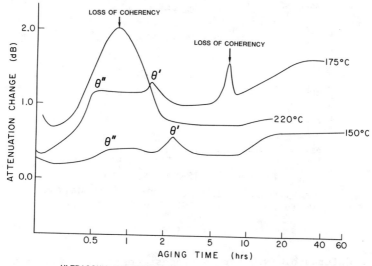

ULTRASONIC ATTENUATION DURING THREE ISOTHERMAL HEAT
TREATMENTS OF 2219 ALUMINUM ALLOY

Figure 6

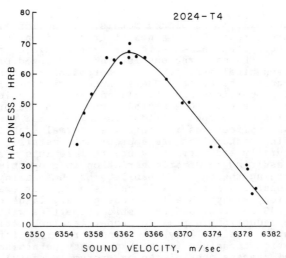

VARIATION OF THE SOUND WAVE VELOCITY
WITH HARDNESS IN AGE HARDENED SAMPLES

Figure 7

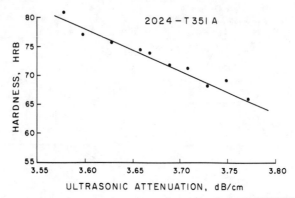

VARIATION OF THE ULTRASONIC ATTENUATION
WITH HARDNESS IN AGE HARDENED SAMPLES

Figure 8

potency of ultrasonic NDC for in-process monitoring and control
of the mechanical properties (hardness, strength) of age-
hardenable aluminum alloys. The correlation suggests that within
a range of thermomechanical treatments, the hardness of the alloy
can be uniquely determined by means of sound velocity and
ultrasonic attenuation.

Nondestructive Charactrerization of Crystallization Processes in
Metallic Glasses

    Amorphous alloys, or glassy metals, are solids with frozen-
in liquid structures. The absence of translational periodicity
in the amorphous state along with the macroscopic compositional
homogeneity are the main reasons for their improved properties,
e.g., high mechanical strength, good corrosion resistance, and
excellent magnetic behavior. Their unusual mechanical, chemical,
and physical properties have stimulated extensive scientific and
technological interest. One serious problem in the processing
and utilization of amorphous alloys that may limit their future
technological applications is their low thermal stability. When
thermomechanical conditions are appropriate, metallic glasses
relax structurally and ultimately crystallize into more stable
structures resulting in drastic variation in properties. The
factors governing the thermal stability of these alloys and their
effect on properties are not well understood. For example, upon
crystallization, amorphous alloys undergo very large changes in
the elastic (40 percent) and anelastic properties with
accompanying reduction in plastic properties (embrittlement).
For this reason, availability of a nondestructive ultrasonic
characterization technique for both property determination and
metallurgical process control can be extremely useful.

    Formation of metallic glasses through rapid solidification
techniques and their subsequent crystallization upon heating has
been the subject of numerous investigations. Characterization of

the range of microstructures thus produced, using nondestructive techniques, is an attractive proposition both in terms of its contribution to the understanding of the kinetic phenomena at play during the various transformations, as well as its ultimate commercial utilization for on-line feedback control of process variables.

Metallic glasses are not thermodynamically stable and tend to structurally relax and finally crystallize upon appropriate heat treatment. Waseda et al. (9) showed that annealing metallic glasses increased the short-range order. Egami (10) used energy-dispersive methods to confirm the log-time kinetics of the relaxation process. His work indicated that the activation energy for relaxation continuously increased with time, which may correspond to the removal of quenched-in defects. The process of crystallization into a stable state involve drastic variations in the properties of the material. Although changes in density and electronic structure associated with the crystallization process are rather minute (11) appreciable variations were observed in several physical properties, including the elastic moduli (12, 14). The crystallization behavior of metallic glasses was examined by several workers (15-18). It was found that small additions of noble metals (Cu, Ag) to Pd-Si greatly stabilize the amorphous phase, and particularly in $Pd_{0.775}Cu_{0.06}Si_{0.165}$ glass (19) which became the subject of numerous investigations. The effect of isothermal annealing on the crystallization kinetics of $Pd_{0.775}Cu_{0.06}Si_{0.165}$ was determined by differential scanning calorimetry (19).

Associated with relaxation and crystallization of metallic glasses variations in the elastic and mechanical properties. Young's modulus E and shear stiffness generally increase by 20-40 percent, but the bulk modulus K increases by only about 7 percent upon crystallization (13). Kursomovic and Scott (20) found that E for $Cu_{60}Zr_{40}$ increased by about 10 percent due to structural relaxation and another 15 percent upon crystallization. In contrast, the density of glassy metals generally changes by only 0.3-1.5 percent.

Straightforward elasticity and wave propagation theories enable one to calculate E(21). For one-dimensional extensional wave propagation in a homogeneous isotropic, linearly elastic solid, Young's modulus is given by $E=v^2\rho$ where v_E is the extensional wave velocity and ρ is density.

Thus the velocity of the extensional waves had to be determined while the ribbons undergo specific heat treatments that induce the crystallization process.

The objectives were as follows:

- Determination of the isothermal transformation kinetics from the amorphous to the crystalline state by means of monitoring changes in sound wave velocity (elastic modulus).

- Calculation of the kinetic parameters: activation energies and crystallite growth regime. Comparison with experimental evidence.

41

    - Information about the relaxational changes occurring in the amorphous state, prior to crystallization.

    - Corroboration with optical and electron microscopy, x-rays and microhardness concerning the relationship between microstructure and properties.

The velocity of the ultrasonic extensional waves was determined by measuring the transit time of a single pulse generated by a laser and detected by a piezoelectric quartz-crystal transducer located at a distance of about 200 mm from the spot on the ribbon irradiated by the laser. The transient load was applied by rapid deposition of energy from a single-pulse of a Q-switched neodymium YAG laser with a wavelength of 1.06 μm. The laser-pulse duration was 15 ns, and the pulse energy, for this specific series of measurements, was about 20 mJ. The laser pulse was line-shaped to produce a nearly plane wave. A photodetector was used to trigger a transient pulse recorder as the laser pulse was generated. Thus, the transit time of the extensional sound wave propagating along the ribbon could be determined to within better than 1 part in 10.

Figure 9 shows the variation of the extensional wave velocity. V_E as a function of crystallization time at three

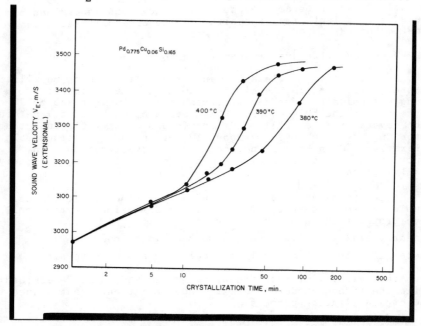

VARIATION OF EXTENSIONAL WAVE VELOCITY IN PdCuSi DURING TRANSITION FROM AMORPHOUS TO CRYSTALLINE STATE AS A FUNCTION OF CRYSTALLIZATION TIME.

Figure 9.

isothermal holding temperatures, 380, 390, and 400°C. The initial sound-wave velocity, in the amorphous state, was found to be 2,970 ms^{-1}, whereas in the fully crystallized state it reached an asymptotic value of 3,490 ms^{-1}. Taking 10.52 and 10.69 g cm^{-3} as the density of the amorphous and crystalline Pd$_{77.5}$Cu$_{16.5}$Si$_{16.5}$ respectively, the Young moduli could be calculated. They were found to be 9.46x10^{3}kg.mm^{-2} for the amorphous state and 13.27x10^3 kg.mm^2 for the crystalline state, i.e., E of about 40 percent. Figure 9 exhibits a relatively sharp increase in the sound velocity during the first five minutes of the isothermal holding. The variation of the extensional wave velocity with crystallization time manifests the dramatic changes in the elastic properties with an increase of about 40 percent in Young's modulus. The sigmoid curves represent the crystallization kinetics which are typical of a thermally activated process. Thus, the kinetic parameters, i.e., activation energy and the growth regime parameters can be determined from the NDC data (2,3). Furthermore, the time dependence of the crystallized volume fraction was found to be compatible with a diffusion controlled growth mechanism (23,24) for a plane boundary growth that is preceded by a rapid nucleation process (25).

Nondestructive Characterization of Microstructurally Modified Surface Layers

High power lasers and electron beams have recently been applied for modifying metal surfaces and improving their physical, chemical, and mechanical properties (7,8). Modification of surface properties may involve incorporation of alloying elements into the surface layer or localized thermomechanical treatment leading to formation of microstructures of desired characteristics. The focused layer or electron beam energy is deposited in such a fashion as to melt a thin layer while the bulk of the material provides the rapid quenching effect.t Typical quenching rates may be of the order of 10^6 Ks^{-1} or higher. Control of the developing microstructures and their characterization is crucial to the understanding of the operating mechanisms responsible for the near-surface reactions and for the exploitation of the potential of this emerging technology for specific applications. Nondestructive characterization may be of prime importance for the on-line, real-time mechanical and physical evaluation of the surface properties, or as basic parameters for engineering design. Demands for safety and quality control have emphasized the need for an acceptable NDE technique for quality assurance and process control. In a recent review (28) various nondestructive evaluation methods were discussed for application on surface coatings:electrical, magnetic, optical, and acoustic. Only those techniques based on acoustic properties showed sufficient promise for engineering applications.

Electron beam glazing has been applied to develop modified metallurgical microstructures such as amorphous layers on crystalline substrates of PdCuSi, deposition of copper on 1100 aluminum samples followed by electron beam melting in an attempt to form an aluminum-copper surface layer on the aluminum bulk, and formation and formation of metastable martensitic microstructures on a pearlitic bulk of AISI 1045 steel (Figs. 10 and 11).

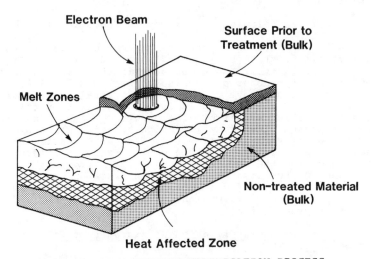

ELECTRON BEAM SURFACE MODIFICATION PROCESS

Figure 10.

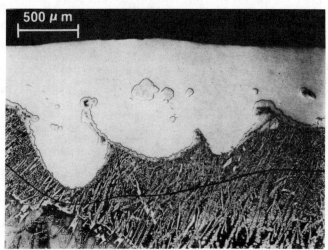

AMORPHOUS PdCuSi LAYER FORMED ON A CRYSTALLINE SUBSTRATE BY ELECTRON BEAM SURFACE TREATMENT

Figure 11.

Ultrasonic NDC of the thermally modified peripheral layer
was determined by means of Rayleigh surface waves, using the
device depicted in Fig. 12. The Rayleigh wave velocity is
frequency independent. However, the extent of penetration of
Rayleigh waves normal to the material surface is frequency

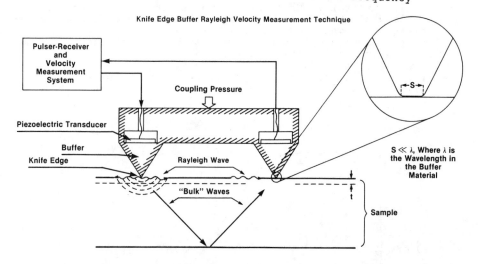

Knife Edge Buffer Rayleigh Velocity Measurement Technique

• The Amplitude of the Rayleigh Wave Decreases Only Due to Attenuation Whereas the
Amplitude of the Cylindrical "Bulk" Wave Decreases as $1/\sqrt{r}$, Where r is the Propagation
Distance.

RAYLEIGH WAVE VELOCITY MEASUREMENTS IN MICROSTRUCTURALLY
MODIFIED SURFACE LAYERS USING THE PIEZOELECTRIC GENERATION AND
DETECTION BY MEANS OF KNIFE EDGE BUFFERS.

Figure 12.

dependent. Frequency analysis of the apparent Rayleigh
velocities enable determination of the elastic properties of the
layer, and the gauging of its average thickness. The Rayleigh
surface wave velocity was measured by means of a wedge device
(Fig. 12) which converts longitudinal waves into Rayleigh waves,
and vice versa. In the case of an amorphous PdCuSi layer, in
which the Rayleigh surface wave velocity is about 20 percent
smaller than the Rayleigh wave velocity in polycrystalline
material, this technique was found to be particularly potent.
Samples of 1053 carbon steel in the pearlitic state were electron
beam heat treated to obtain a microstructurally modified surface
layer of martensite. The Rayleigh velocity of the uniform
pearlite was found to be higher by more than 2 percent than that
of martensite. By measuring the Rayleigh wave velocity over a
frequency range, thus varying the penetration depth of the
Rayleigh surface waves, the modified layer depth could be
nondestructively determined (Fig. 13). The Rayleigh velocity
remains constant as long as the Rayleigh waves probe a uniform
material. When the Rayleigh waves, due to decreasing frequency
or increasing wavelength, begin to sample simultaneously both the

modified martensite layer and the pearlite substrate, the measured Rayleigh velocity is expected to increase. This behavior is apparent in Figure 13. The lower curve should increase and merge asymptotically into the upper curve (straight line) that exhibits the constant Rayleigh velocity in the pearlitic substrate. These measurements demonstrate the feasibility of the technique to nondestructively characterize the properties of a modified surface layer, and to evaluate its thickness. Preliminary studies have shown that the laser generation and laser-interferometric detection of Rayleigh surface waves can be applied for nondestructive, non-contact, evaluation of surface layers. Development of an in-process, real-time method would be of great technical importance.

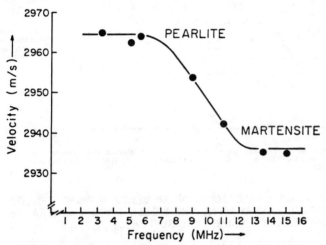

VARIATION OF THE APPARENT RAYLEIGH WAVE VELOCITY WITH FREQUENCY
(RECIPROCAL OF DEPTH OF PENETRATION) IN AISI 1045 STEEL.
REGIONS OF CONSTANT VELOCITY (e.g. MARTENSITE) OCCUR WHEN THE
RAYLEIGH WAVES PROBE UNIFORM LAYERS OF MATERIAL.

Figure 13.

References

1. M. Rosen, S. Fick, R. Reno, E. Horowitz and R. Mehrabian, Kinetics of precipitation hardening process in 2219 aluminium alloy studied by sound veloctiy and ultraosnic attenuation, Mater. Sc. Eng. 53, 1982, p. 163.

2. M. Rosen, H.N.G. Wadley and R. Mehrabian, Crystallization kinetics study of amorphous PdCuSi by ultrasonic measurements, Scripta Metall., 15, 1981, p. 1231.

3. J. Chang, F. Nadeau, M. Rosen and R. Mehrabian,
    Crystallization kinetics study of amaorphous $Zr_{50}Cu_{50}$
    by means of ultrasonic measurements and microhardness,
    Scripta Metall., 16, 1982, p. 1073.

4. F. Nadeau, M. Rosen and R. Mehrabian, Effects of
    composition on the elastic properties of amorphous and
    crystalline Cu-Zr alloys, Johns Hopkins University
    Report CMR-NDE-7, 1982.

5. M.J. McSkimin, J. Acoust. Soc. Amer., 33, 1961, p. 12.

6. C.H. Palmer, S. Fick and M. Rosen, a dual-probe laser
    interferometer for contactless determination of
    ultrasonic wave velocities and attenuation, Johns
    Hopkins University Report CMR-NDE-11, 1983.

7. E.M. Breinan, and B.K. Kear, Rapid solidification laser
    processing of materials for control of microstructures and
    properties, In Proceedings of the 1st International
    Conference on Rapid Solidification Processing, Reston, VA,
    1977, R. Mehrabian, B.J. Kear and and M. Cohen, Eds.,
    Claitors Publ. I, 1978, pp. 87-103.

8. T.R. Tucker and J.D. Ayers, Surface microstructures-
    produced by scanning electron beams, In Proceedings of
    the 2nd International Conference on Rapid
    Solidification Processing, Reston, VA, R. Mehrabian,
    B.K. Kear and M. Cohen, Eds., Claitors Publ. II, 1981,
    pp. 206-211.

9. Y. Waseda, H. Okazaki and T. Masumoto, J. Mat. Sc., 12,
    1977.

10. T. Egami and T. Ichikawa, Mat. Sc. Eng., 32, 1978, p.
    293.

11. H.S. Chen and J.T. Krause, Scripta Met., 11, 1977, p.
    761.

12. H.S. Chen, M.J. Leamy and M. Barmatz, J. Noncrystalline
    Solids, 5, 1971, p. 444.

13. B. Golding, B.C. Bagley and F.S.L. Hsu, Phpys. Rev.
    Lett., 29, 1972, p. 68.

14. H.S. Chen, J. Appl. Phys., 49, 1978, p. 3289.

15. D. Duwez, R.H. Willens and R.1L. Crewdson, J. Appl.
    Phys., 36, 1965, p. 2267.

16. J.J. Burton and R.P. Ray, J. Non-Crystalline Solids, 6,
    1971, p. 393.

17. T. Matsumoto and R. Maddin, Acta Met.j, 19, 1971, p.
    725.

18. B.G. Bagley and E.M. Vodel, J. Non Crystalline Solids,
    29, 1975, p. 18.

19. H.S. Chen and D. Turnbull, Acta Met., $\underline{17}$, 1969, p. 1021.

20. A. Kursumovic and M.G. Scott, Appl. Phys. Lett., $\underline{37}$, 1980, p. 620.

21. K.F. Graf, Wave motion in elastic solids, Ohio State University Press, Columbus, OH, 1975.

22. H.A. Davies, Rapidly Quenched Metals III, I, P. 1 Proc. Third Internat. Conf. Rapidly Quenched Metals, The Metals Society, London, England, 1978.

23. W.A. Johnson and R.F. Mehl, Trans. AIME, $\underline{135}$, 1939, p. 416.

24. J.W. Christian, Physical Metall., R.W. Cahn, Ed., North Holland Press, Amsterdam, London, 1970, p. 471.

25. J.W. Cahn, Acta Metall., $\underline{4}$, 1956, p. 449.

26. H.S. Chen, Rep. Prog. Phys., $\underline{43}$, 1980, p. 355.

27. D. Weaire, M.F. Ashby, J. Logan and M.S. Weins, Acta Metallurg., $\underline{19}$, 1979, p. 799.

28. I.A. Bucklow, Proceeding sof the International Conference on the Advances in Surface Coating Technology, London, 1978, p. 61.

ULTRASONIC SENSOR FOR CONTROL OF SURFACE MODIFICATION

B.J. Elkind, M.Rosen
Center for Nondestructive Evaluation
Materials Science and Engineering Department
The Johns Hopkins University
Baltimore, Maryland 21218

H.N.G. Wadley
National Bureau of Standards
Gaithersburg, Maryland 20899

Abstract

A nondestructive method is presented for the determination of depth and hardness of modified surface layers. This method is based upon the differences in the elastic constants and density between a modified surface layer and the substrate beneath it. These parameters can be conveniently observed as variations in velocity with changes in frequency of Rayleigh surface waves. As a test of this technique, studies were conducted on an AISI 1053 plain carbon steel that has been subjected to electron-beam irradiation. The electron-beam treatment allowed for the creation of thin, microstructurally-modified surface layers. The (frequency-independent) Rayleigh surface wave velocity of homogeneous pearlitic and martensitic samples of this steel were 3002 m/s and 2960 m/s, respectively. The lower velocities of the martensite resulted from a softening of the elastic constants. On samples with an approximate 1 mm rapidly solidified, martensitic surface layer on a pearlitic substrate, the Rayleigh velocity varied form 2984 m/s at low frequency (deep penetration into the substrate) to 2960 m/s at high frequency (penetration confined only to the surface layer). A rapid increase in velocity occurred as the depth of penetration extended beyond the depth of the modified surface layer and an empirical method for determining penetration depth (hence, layer thickness) to within 10%.

INTRODUCTION

Microcrystalline modified surface layers may possess enhanced resistance to erosion, wear and corrosion. These beneficial attributes are important in a diversity of high technology applications where material performance has been the primary limitation. If this material's processing approach is to be utilized in practice, it is important to control spatial nonuniformities (of microstructure and chemistry), porosity, cracking, depth of modification and surface topology. Microstructural and chemical heterogeneity are believed to be caused by uncontrolled variations in process variables; principally temperature gradient, solidification rate, cooling rate, and by convective mixing in the melt. Microstructural inhomogeneities may also be caused by the annealing of already processed material by subsequent melt passes in adjacent material.

The formation of amorphous surface layers or the layer glazing process, as it is sometimes referred to, has been demonstrated to be technically feasible, but its widespread application is limited by two problems:the difficulty of producing deep (>20 μm) amorphous layers and the elimination of cracking. The former problem arises either from the reduction of temperature gradient for deeply melted layers, which causes a reduction of solidification velocity to a value below the threshold for glass formation[1] or from the occurrence of amorphous to crystalline transitions in the heat affected zones of previously vitrified material during sequential melting passes in adjacent material. The cracking problem, on the other hand, arises because of the large tensile residual stresses that are produced by solidification shrinkage strains.

The work reported here concerns development of an ultrasonic technique to characterize the depth of surface modified microcrystalline layers and their hardness. Clearly, if the potential of surface modification with directed high energy sources is to be fully realized, techniques (or sensors) need to be developed to measure, in process, key process variables (those controlling microstructure) and to detect cracking and other deleterious conditions. With such a capability it may then be possible to use automated process control[2] as a means for reliably producing uniformly high quality material. Previous research[3] has shown the potential of acoustic emission for detecting cracking during surface modification.

The basis of the technique stems from the difference in ultrasonic velocity between the layer and substrate. In this study, the technique has been applied to the characterization of a surface modified layer on steel substrate. Since the physical basis of the technique is the difference in ultrasonic velocity of substrate and layer, it should be noted that it is actually more pronounced if the surface layer is amorphous, because the difference in elastic constants, and thus ultrasonic velocity, of the (crystalline) substrate and the (amorphous) layer is larger. Consequently, the microcrystalline layer study is in some respects a worst case test of the approach.

When applied to steels, electron beam surface modification has been observed to cause substantial improvements in hardness due to the formation of martensite/bainite and carbide precipitate refinement.[4] Systematic investigations have shown that the hardness and depth of modification are controlled by scanning rate.[5-7] Thus, variation of the scan velocity could provide a convenient potential method for the control of hardness and melt depth, provided they can be measured in-process. The use of ultrasonic surface acoustic waves (Rayleigh waves) has been investigated here for the characterization of the depth and hardness of electron beam modified surface layers.

For NDE surface layer characterization, the technique utilized was the ultrasonic generation of Rayleigh surface waves. These waves can be induced in a material by means of piezoelectric generation and detection with standard ultrasonic transducer. The surface wave energy diminishes rapidly with increasing depth into the bulk of a material. The rate of decrease in amplitude is, however, dependent upon wavelength; at a depth of approximately one wavelength, the particle amplitude is only about 20% (and the particle energy is only 4%) of that at the surface. Therefore, this waveform represents a two-dimensional wave that attenuates as $1/r$ (r is the distance from point source). In addition, Rayleigh wave propagation in a uniform material is nondispersive in that Rayleigh velocity is independent of frequency. The velocity is dependent only upon density and elastic constants in homogeneous media.

The velocity (V_R) of Rayleigh surface waves can be expressed in terms of a constant (γ), and the shear wave velocity and Poisson's ratio (ν):

$$\gamma = V_R/V_t = \frac{0.87+1.12\nu}{1 + \nu}$$

Eq. #1

Rayleigh wave fronts are elliptical in nature and travel along the direction of propagation in a planar fashion and orthogonal to the surface. This wave motion is illustrative of the fact that the Rayleigh wave is a combination of vertically polarized shear (SV) waves and longitudinal waves. The major axes of these ellipses depict the vertical particle displacements (SV motion) of the Rayleigh waves, whereas the minor axes represent the horizontal (or longitudinal) displacements. For surface layer analysis, the orthogonal component of the Rayleigh wave motion is of principal interest and the effective depth of penetration of the Rayleigh surface wave may be assumed to be approximately one wavelength (λ_R).

penetration depth of the Rayleigh waves may be varied by changing the frequency of the waves of constant velocity (in an homogeneous medium) according to the relation: $V = f\lambda$ where V_R – Rayleigh surface wave velocity (phase), f is the

Rayleigh wave frequency and λ is the Rayleigh wavelength, which can be assumed to be equal to the penetration depth. Therefore, the measurement of velocity dispersion could potentially provide rapid, non-destructive method for selectively probing sub-surface properties[8-10].

EXPERIMENTAL

Specimen Preparation and Characterization

Several 7.6cm x 2.5cm X 0.6 cm slabs of AISI 1053 (0.53% carbon) steel were austenitized at 1025° c for 45 minutes. One of the samples was subsequently quenched in an ice-brine to produce a bulk martensite. The remaining samples were furnace cooled to to 650° C, allowed to isothermally transform for 15 minutes and air-cooled to room temperature. This resulted in a pearlitic microstructure. The specimens were eventually subjected to electron-beam surface

heat-treatments. The electron-beam glazing technique employs a
high intensity/energy collimated beam of electrons. The
directed energy beam strikes the surface of metal or alloy,
causing the subsequent heat-treatment (or melting) of a surface
layer. High cooling rates on the

order of 10^6 K/s are obtained by self-substrate quenching. The
parameters that dictate the effectiveness of this process are
the following: beam energy, beam intensity, beam shape, and scan
rate.

After completion of the preliminary characterization by
means of optical metallography and mechanical testing, each of
the samples was subjected to an identical surface modification
process. A 25 KV accelerating voltage was employed to
accelerate a 2.5 mm diameter beam of electrons (emitted from a
tantalum filament) to the specimen surface. The beam current
was 9 milli-amps. The beam power was, therefore, 225 watts.
The scan rate of the electron-beam across the length of each
specimen was 0.3 cm/s. After each scan of the electron-beam,
the beam was translated 0.0254 cm across the width of the
specimen, so as to microstructurally modify the entire surface.
After the initial heat treatments of the specimens to obtain
bulk pearlitic and martensitic specimens, several preliminary
measurements were performed. Rockwell macrohardness tests were
conducted and the densities of the specimens were measured. In
addition, standard metallography and microstructural analyses
were carried out. Longitudinal and shear ultrasonic bulk wave
velocity measurements were obtained on these samples, utilizing
the pulse-echo overlap technique. Elastic moduli were
calculated from the ultrasonic data.

Ultrasonic Rayleigh Wave Characterization

In the present investigation, the generation of Rayleigh
waves is obtained by means of a specially designed and
constructed mode-conversion device. An illustration of this
device appears in Figure 1. The wedge device consists of two
machined 2024 aluminum wedges that are placed in an aluminum
sliding mount. Each wedge has an active surface area in contact
with the specimen of approximately 2.54cm long by 700 m wide
(w). A piezoelectric transducer is mounted inside each wedge;
the wedges are machined down to

knife-edge and the sample is approximately 60°. The generation
and detection of the Rayleigh waves are accomplished by
conversion of the generated longitudinal wave and by subsequent
reconversion of the detected wave.

Rayleigh wave frequencies from 2.2 to 15.5 MHz were used to
characterize electron-beam treated samples. The minimum value
of frequency was determined by the need to have at least a half-
wavelength equal to (or less than) the width of the wedge i.e.,

$$< \quad F_R = V_R / 2w = \frac{3002 \text{ m/s}}{1400 \text{ μm}} = 2.1 \text{ MHz} \qquad \text{Eq. #2}$$

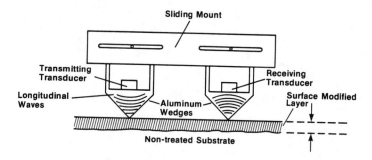

Rayleigh Wave Mode Conversion Device

Figure 1.   SCHEMATIC DIAGRAM OF THE MODE CONVERSION TECHNIQUE
FOR GENERATION AND DETECTION OF RAYLEIGH SURFACE
WAVES ON MODIFIED SURFACES.

At the lower frequency limit, waveguide effects
predominate, thereby prohibiting low frequency mode conversion
coupling of longitudinal waves into Rayleigh waves to the
sample. The observed frequency minimum of 2.2 MHz correlates
well with the theoretically predicted frequency minimum of 2.1
MHz. At high frequencies, difficulties arise from enhanced
attenuation, which increases with frequency. A likely source of
this is grain boundary scattering [11].

RESULTS

Homogeneous Samples

The results of preliminary characterization of homogeneous
pearlitic and martensitic microstructure specimens are shown in
Table 1. Noteworthy, in Table 1 is the observation that both
the longitudinal and shear wave velocities of pearlite are
greater than those of martensite, an observation that has been
reported by others. [12]

For isotropic linear elastic materials, the density and
wavespeeds can be used to determine Young's modulus, the

bulk and shear moduli and Poisson's ratio.[9] Furthermore, it
is possible to predict the Rayleigh wave velocity, V_R,

using the relation expressed in Eq. 1. These physical
quantities are presented in Table 2.

53

Table 1

MEASURED PHYSICAL PROPERTIES OF HOMOGENEOUS AISI 1053
SAMPLES

PHYSICAL QUANTITY	PEARLITE	MARTENSITE
Density	7796 $Kg.m^{-3}$	7779 $Kg.m^{-3}$
Shear Wave Velocity	3239 $m.s^{-1}$	3167 $m.s^{-1}$
Longitudinal Wave Velocity	6011 $m.s^{-1}$	5945 $m.s^{-1}$
Rockwell Macro-Hardness	R_B = 91.5	R_C = 57.0

Table 2

— DEDUCED PHYSICAL PROPERTIES OF HOMOGENEOUS AISI 1053
SAMPLES

PHYSICAL QUANTITY	PEARLITE	MARTENSITE
Young's Modulus (E)	211.9 GPa	203.1 GPa
Bulk Modulus (K)	172.7 GPa	170.9 GPa
Shear Modulus (G)	81.78 GPa	78.00 GPa
Poisson's Ratio(ν)	0.295	0.301
Rayleigh wave velocity (V_R)	3002 $m.s^{-1}$	2939 $m.s^{-1}$

Microstructure of Surface Modified Material

The surface-modified samples (sectioned perpendicular to
the electron beam pass direction), polished and etched in 2%
nital to reveal the microstructures produced and to facilitate
microhardness measurements. Figure 2 shows, at low
magnification, a view of a region about midway across the width
of AISI 1053 modified material shown together with a
microhardness depth profile. Starting at the bottom, it can be
seen that the substrate had a pearlite microstructure with a

grain size of 25 μm, (Figure 2), the peak temperature
experienced during modification increases. This at first
results in dissolution of cementite, then a transformation to
austenite and finally melting during the heating part of the
thermal cycle. Rapid cooling from these states then produces a
wide spectrum of microstructures that are vertically separated.
Subsequent melt passes further complicate the microstructures
due to tempering in the heat affected zone. Results of
microhardness and microstructural analysis of modified AISI 1053
are summarized in Table #3.

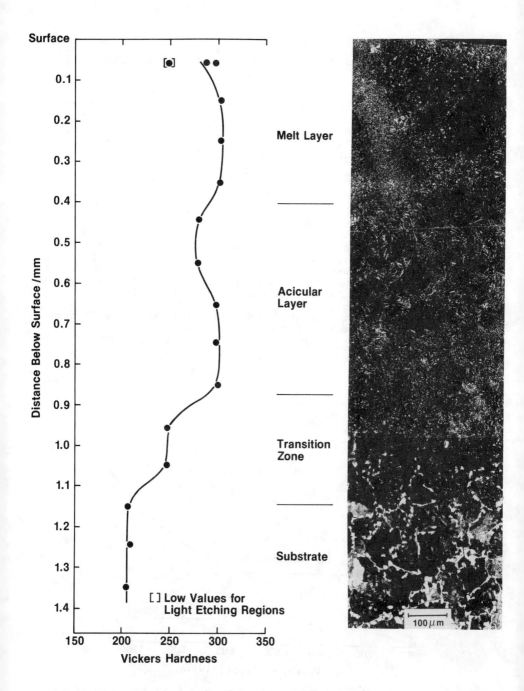

Figure 2.   A NARROW REGION OF THE SAMPLE REVEALING THE VERTICAL
SEPARATION OF MICRO-STRUCTURE (2% NITAL ETCH)
TOGETHER WITH THE MICRO-HARDNESS VARIATION WITH DEPTH

Table 3

PROPERTIES OF SURFACE MODIFIED AISI 1053 ALLOY

REGION	THICKNESS	MICROSTRUCTURE	HARDNESS
Melt Layer	0.330 mm	Tempered Upper Bainite	R_c=22-29
Circular Layer	0.435 mm	Bainite	R_c=27-30
Transition Zone	0.300 mm	Pearlite/Spheroidal Carbides	R_c=22-29
Substrate	5.2 mm	Pearlite	R_b=94.0

Rayleigh Wave Measurements

Figure 3 shows a plot of the measured Rayleigh wave velocity as a function of wavelength (penetration depth) for the surface modified layer and the unmodified pearlitic substrate as obtained from the unmodified reverse side of the sample. It can be seen that the substrate exhibits a constant velocity independent of wavelength with an average value of 3001 m/s in excellent agreement with the predicted value (in Table 2) from bulk wave measurements. The absence of dispersion from a homogeneous sample indicates that the experimental approach does

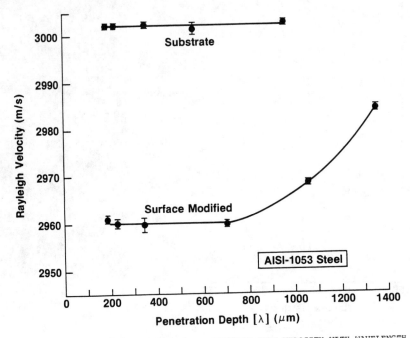

Figure 3. VARIATION OF THE RAYLEIGH WAVE VELOCITY WITH WAVELENGTH (PENETRATION DEPTH) FOR AISI 1053 STEEL.

not introduce any significant velocity measurement errors. The
velocity measured on the surface modified layer exhibits
significant dispersion. At high frequencies (shallow
penetration depths) a limiting velocity of 2960 m/s is achieved.
This represents the velocity of the layer and is   20 m/s
greater than the value predicted from bulk wave measurements on
a homogeneous martensite sample. This difference is most
probably due to the bainite microstructure (Figure #4) of the
layers since bainites are well known to have elastic properties
intermediate between those of pearlite and martensite.[12] We
see in Figure 3 that as the wavelength increased the velocity
began to approach the value of pearlite consistent with an
enhanced contribution of the substrate to the elastic constants
sampled by the wave..

Figure 4.   BAINITIC MICROSTRUCTURE INDICATIVE OF THE
            MODIFIED SURFACE LAYER IN AISI 1053 (2% NITAL
            ETCH @ ≃ 1250 x MAG.).

DISCUSSION

     Rayleigh waves propagating over the surface of electron-
beam modified surfaces, sample only the modified (martensitic)
layer at high frequencies (since the wave does not penetrate the
layer) and some combination of the layer and substrate
(pearlitic) at lower frequencies. In the high frequency region,
the velocity is found to be approximately frequency independent,
consistent with wave propagation (theory) in an approximately
uniform, non-dispersive medium. Then the limiting velocity can
be viewed as a characteristic of the layer microstructure and
not its depth. The velocity for AISI 1053 surface layers is
intermediate between that anticipated for martensite and
pearlite, and is consistent with the observed bainite state.

There thus appears considerable merit to the use of this asymptomatic value to characterize layer hardness.

Rayleigh waves with greater penetration depths yielded velocities that were intermediate between those of the substrate and the layer. Provided the substrate is much thicker than the layer, then when the wavelength (penetration) is very much greater than the layer thickness, the velocity approaches the Rayleigh velocity for pearlite because contributions from the layer become increasingly small. This limit was not reached in these samples due to their finite thickness, but sufficient data for a depth determination were obtained nevertheless.

To determine the layer thickness, the region of interest is the intermediate velocity region.,1, where the velocity results from the simultaneous sampling of several different microstructures. For example, in AISI 1053 at a frequency of 4.2 MHz (penetration depth = 705 m), the Rayleigh velocity was 2960 m/s; however at 2.2 MHz (penetration depth = 1357 m), the apparent Rayleigh velocity was 2984 m/s. The velocity thus depends upon both the characteristic velocities of the layer and substrate and the layer depth.

The inverse problem of deducing the layer properties from velocity despersion data is complex and is the object of ongoing study. Here, we use a very simple analysis that gives results of reasonable accuracy. For the case where the wave penetrates the surface modified layer, we may, to a first approximation, assume the apparent velocity to be determined by the layer thickness: wavelength ratio according to a law of mixtures:

$$V = XV_L + (1-X)V_S \qquad \text{Eq. \#3}$$

where V_L = the velocity of the layer

$\qquad V_S$ = the velocity of the substrate

$\qquad x = x_L/\lambda$, the ratio of layer thickness to wavelength
$\qquad\qquad$ (n.b. we define (x = 1 for λ X_L)

In the long wavelength limit, x tends to zero and the apparent velocity is V_S, while for λ X_L, the velocity

becomes that of the layer, V_L.

To determine the layer depth we rearrange the equation above:

$$X_L = \frac{(V-V_S) \,\lambda}{V_L - V_S} \qquad \text{Eq. \#4}$$

Substituting data from Figure 3 into this expression yields a value for a modified layer depth of 600-800 μ m. The variability depends upon the precise value of wavelength taken. Metallography for AISI 1053 indicated that the combined circular and melt depth was 850 μ m, in reasonable agreement to that deduced bearing in mind the simplicity of a law of mixture model.

A more complete analysis incorporating exponential weighting of the Rayleigh wave particle displacement, might give a more exact inversion, and thus deeper insight, into the functional behavior of the modulus with depth. However, for the present purpose, this simple approach appears sufficient, at least for the AISI 1053 alloy. Indeed, an even simpler depth determination criterion such as the penetration depth at which the velocity curve increases 2 m.s^{-1} above the short wavelength velocity limit predicts the metallographically deduced depths of hardening for both AISI 1053 and 1044 alloys that have been subjected to surface modification to within $+25\mu$ m.

The Rayleigh wave technique has proved, therefore, to be an accurate and reproducible technique for determining the depth of hardening of surface-modified steels. Since it is nondestructive, and methodologies which would not interfere with processing are emerging, it can potentially be used for on-line quality control purposes. If account is taken of the effect of sample temperature fluctuations on the velocity, it could serve the role of a process control sensor for feedback control of depth of modification and hardness, particularly if emerging noncontact measurement methodologies are utilized.

SUMMARY

. Samples of pearlitic plain carbon steels were subjected to an electron-beam glazing process resulting in a (case-hardened) microstructurally modified surface layer.

. Hardness and ultrasonic velocity data indicated that the modified surface layers possessed a higher hardness but a lower velocity than the substrate materials. This decrease in velocity may be a good indication of the level of hardness attained in the modified layer.

. Dispersion of the Rayleigh wave velocity can be utilized to make a nondestructive estimate of the depth of the microstructurally modified surface layer.

ACKNOWLEDGEMENTS

We are grateful to Dr. R. Schaefer for the assistance and advice with electron beam surface modification, to C. Brady for help with metallography and to the Defense Advanced Research Projects Agency for funding of this work under DARPA Order Number 4275 (Major S. Wax, Program Monitor).

REFERENCES

1. W.J. Boettinger, F.S. Biancaniello, G.M. Kalonji and J.W. Cahn, Rapid Solidification Processing: Principles and Technologies II, M. Cohen, B.H. Kear, and R. Mehrabian, eds. Claitor's Press, Baton Rouge, LA, 1980, p. 50.

2. R. Mehrabian and H.N.G. Wadley, Journal of Metals, February 1985, p. 51.

3. R.B. Clough, H.N.G. Wadley, and R. Mehrabian, Lasers in Materials Processing, Ed., E.A. Metzbower, ASM, Ohio, 1983, p. 37.

4. B.G. Lewis, D.A. Gilbert, P.R. Strutt, "Rapid Solidification Processing: Principles and Technologies II, "Eds., Cohen, Kear and Mehrabian, Claitors Publishing Division, Baton Rouge, 1980, p. 221.

5. J.E. Jenkings, Thin Solid Films, 68, 1981, p. 341.

6. Ibid, p. 343

7. J. Masumder, Journal of Metals, 1983, p. 18.

8. G.J. Curtis, Ultrasonic Testing, Ed., J. Szilard, 1982, Wiley, New York, p. 297.

9. W.L. Plant, Elastic Waves in the Earth, 1979, Elsevier Scientific Publ. Co., New York, p. 102.

10. I.A. Viktorov, Rayleigh and Lamb Waves, 1967, Plenum Press, New York.

11. E. Fisher, D. Zwick, D. Van Hook, G. Henklemen, B. Smith, "Application of Metal Single Crystal Wedges to High Frequency Rayleigh Wave Propagation<" Journal Nondestructive Evaluation, Vol. 2-3, p. 231, (1982).

12. N. Grayell, D.B. Ilic, F. Stanke, C.H. Chour and T.C. Shyne, Proc. 1979 Ultrasonic Symposium, IEEE, 1979, p. 273.

REAL-TIME MONITORING OF MICROSTRUCTURAL TRANSFORMATIONS USING

SYNCHROTRON AND FLASH X-RAY DIFFRACTION

John M. Winter, Jr. and Robert E. Green, Jr.

Center for Nondestructive Evaluation
The Johns Hopkins University
Baltimore, MD., 21218

Abstract

The use of x-ray diffraction to monitor dynamic changes in crystal structure has long been recognized as a uniquely powerful analytical tool. Such studies require obtaining x-ray diffraction records within relatively short exposure times in order to study rapid structural alterations. Depending upon the requirements of the experiment, these records may be either photographic or electro-optic images. One technique described uses an x-ray system composed of a pulsed x-ray generator driving a cold cathode field emission tube, and an electro-optic imaging capability. The merits of various options for the imaging capability are examined. A second technique, which utilizes the exceptionally high photon flux of a synchrotron to shorten exposure times, is discussed.

Introduction

The use of x-ray diffraction to monitor dynamic changes in crystal structure generally requires the use of some form of gain or amplification to enhance the intensity of the image generated by the diffraction process. The following considerations examine the general design approaches which have been used, as well as specific examples of applications to various real time x-ray diffraction imaging problems.

Generic System Configurations

Figure 1 shows the two generic systems configurations which have been employed (1). The first is the direct method, using either a rotating anode tube or synchrotron radiation as an x-ray photon source. From such a high intensity source, one obtains a high intensity diffraction image, permitting use of a low gain, high resolution electro-optical imaging system. As also shown in Fig. 1, the second is the indirect method, which uses a conventional x-ray tube source. With this method, a lower intensity source leads to a low intensity diffraction image, so one needs a high gain electro-optical imaging system. A variety of combinations can be used, but they all share the disadvantage that the resolution always is limited by the fluorescent screen used to convert x-ray photons to visible light photons. However, all such systems share the advantage that they are the least susceptible to radiation damage.

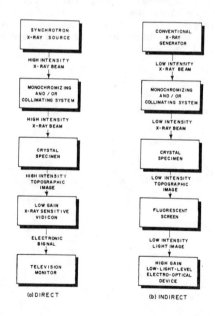

Figure 1 - Block diagram of the two basic electro-optical methods for rapid viewing and recording of x-ray images.

Implementation of the Indirect Configuration

Most of the work of the present authors has been done using the indirect configuration. It is well suited to a modular design approach, which in turn is generally the least expensive, and is relatively easy to update with minimum effort as new developments are made. Figure 2 shows two image intensifier tube designs typical of those available for use in the indirect configuration (2). Figure 2(a) depicts a first generation tube. It is made of three electrostatically focussed tubes, each fiber-optically coupled to the next, with a total light gain of one or two million. (Usually a few hundred thousand is sufficient gain to obtain a transmission x-ray diffraction image with conventional machines, as long as the specimen's thickness times its linear mass absorption coefficient is the order of one.) Figure 2(b) shows a second generation tube, a development which is not really so new, but it has only relatively recently become commercially available. It employs a microchannel plate (MCP) which results in a smaller overall physical size for the tube than the first generation type. Photon gain is typically a few hundred thousand. (With a synchrotron radiation source, one is willing to sacrifice some gain in exchange for compact size.) Some image tubes have two MCP's, with a resulting gain comparable to the three stage first generation tubes.

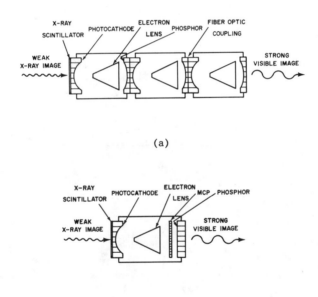

(a)

(b)

Figure 2 - Schematics of (a) first generation image intensifier tube, and (b) second generation image intensifier tube.

To complete the indirect system, the output of the intensifier tube is then coupled in some way to a means of dynamically recording the images at the intensifier output. Originally, this frequently took the form of a 16mm camera lens coupled to the output screen of the intensifier tube. In more recent times, the movie camera has been replaced with a TV camera and VCR.

And in more sophisticated configurations, the lens coupling has been replaced with various coherent fiber optic direct coupling arrangements. As mentioned above, the fluorescent screen used to convert the x-ray photons to visible light photons is still the limiting factor in spatial resolution in any such system.

One example of the use of an indirect configuration is illustrated in Fig. 3 which shows a schematic diagram of an arrangement used in the authors' laboratory to record in real time the x-ray diffraction images from ribbons of chill block melt spun alloys. A small ribbon of alloy is rapidly solidified on a rotating chill wheel, ejected from the wheel at a velocity of about 100 ft/sec, and x-rayed within a few inches of where it underwent solidification. The result is a continuous record of diffraction images within one half to one millisecond after solidification. The goal of the experiment was to compare this immediate data with subsequent diffraction images taken hours or days after solidification to document a "time constant for crystallization" as a function of storage temperature for the alloys under study.

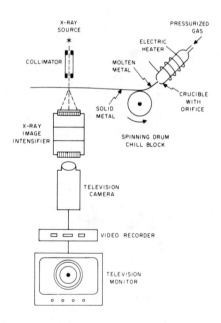

Figure 3 - Schematic of arrangement for real time
imaging of diffraction patterns from ribbons
of chill block melt spun alloys.

Real Time X-Ray Topography

X-ray topography is one particular branch of x-ray diffraction which can derive many benefits from real time imaging. Several examples as adapted to specific techniques of topography are considered in this section. All topographic methods are designed to examine the defect structure in single crystal material, usually by magnifying the image obtained by classical Laue diffraction from a single crystal. One arranges to illuminate a relatively large area of the single crystal under study either by using an expanded x-ray beam or by moving the specimen across in front of the beam.

Figure 4 shows a schematic of one particular method of x-ray topography known as the Lang method. The Lang method requires that the film (or other imaging detector) move in synchronization with the specimen. Figure 5 shows the first topograph which combined the Lang technique with real time image intensifier hardware (3). The specimen was diamond. The image shown in Fig. 5(a) was recorded on a nuclear emulsion plate after an eleven hour exposure, and the image shown in Fig. 5(b) was the essentially instantaneous output of an image intensifier system.

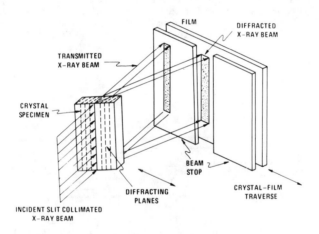

Figure 4 - Schematic of the Lang method of topography.

Figure 6 is a schematic of a system using the direct configuration as described by Chikawa (4). It employs the Lang technique, so both crystal and detector must scan. Since the detector is a TV camera, the scanning device needs to be quite sturdy. As with any direct configuration system, a necessity of this system is a very powerful x-ray source. Originally Chikawa used a unit which could deliver 500 ma at 60 KV, and more recently, one which delivers up to 1000 ma. Even with these intense sources, the imaging system

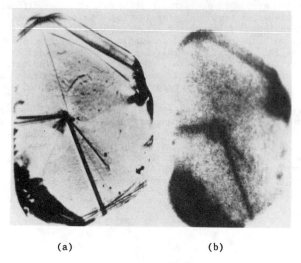

(a)   (b)

Figure 5 - Topographs of natural diamond using a 12 sec specimen
traverse; (a) recorded on 50 micron thick Ilford L4 nuclear
emulsion, 11 hour exposure, (b) Polaroid photograph of
television screen displaying image intensifier output.

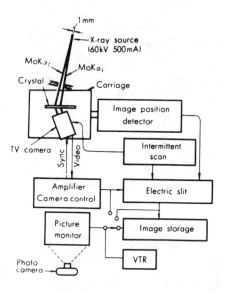

Figure 6 - Schematic of direct X-ray topography
video display system.

is insensitive, so the data are twice integrated. First, storage at the detector is accomplished by cutting off the electron beam which scans the vidicon target. The storage time is limited by the rate of lateral charge diffusion in the vidicon target. Second, the image is stored in a computer, and subsequently synthesized. After a minute or so, this doubly integrated image is displayed on a TV monitor and photographed. (The electric slit serves the same purpose as a beam stop.) Figure 7 shows image of dislocations in a silicon single crystal wafer as obtained by Chikawa (5). Figure 7(a) is the image on a nuclear emulsion plate, Fig. 7(b) is the synthesized video image obtained on a TV monitor after integration in time, Fig. 7(c) shows what one sees in the instantaneous area of electric slit, and Fig. 7(d) shows a line scan of the video waveform along the dashed line in Fig 7(c).

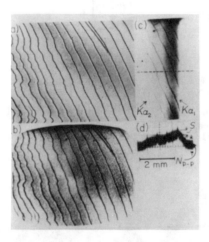

Figure 7 - Topographic images of static dislocations in a silicon crystal; (a) Image on a nuclear emulsion plate, (b) synthesized video image obtained on a TV monitor after integration in time, (c) instantaneous view of area in electric slit, (d) line scan of the video waveform along the dashed line in (c).

Another example is shown in Fig. 8, which is a scheme used by Boettinger, Burdette, Kuriyama, and Green for topographs using a first generation three stage image tube and the indirect method (6). The figure shows the general optical arrangement for what is called asymmetric crystal topography (ACT). In this method of real time topography, an asymmetric cut crystal is used to expand the cross-sectional size of the x-ray beam. Figure 9 shows some surface relection topographs of a copper crystal with several subgrains (6). This is a series of photos taken from the output phosphor of the image intensifier using a 35 mm camera. Figures 9(a), 9(b), and 9(d) correspond to optimum glancing angles to bring out specific subgrains. Figure 9(c) is taken at the best glancing angle for general coverage.

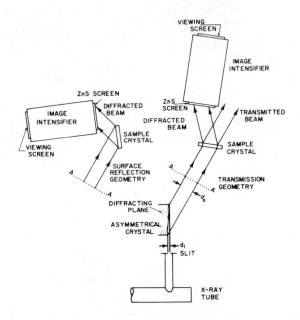

Figure 8 - Optical alignment of asymmetric crystal topography (ACT) showing positions of first and second crystals and X-ray image intensifier in surface reflection and transmission geometries.

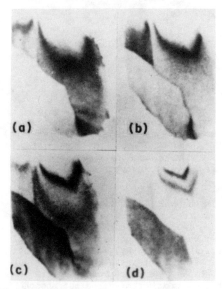

Figure 9 - Photographs of image produced by ACT surface reflection topographs of a copper crystal. Glancing angles in (a), (b), and (d) are optimum for specific subgrains, (c) is best overall orientation to indicate all subgrains.

Some of the capabilities for dynamic topography facilitated by the high flux available from the National Synchrotron Light Source (NSLS) at Brookhaven National Laboratory will be considered next. An example is yet another type of topography, known as white beam transmission topography (WBTT). The technique is similar to Lang topography, except one doesn't scan the specimen and film. Instead, in many cases, the incident beam is large enough in cross-section to illuminate the entire crystal at the same time. In addition, the flux is sufficient to permit moving back far enough from the specimen to get spatial separation of a large number of Laue images. One may choose to record a whole array of diffracted Laue images in one large area exposure. Or an individual Laue image may be singled out and magnified. (Each image corresponds to diffraction from one particular set of lattice planes.) Figure 10 shows the result of a WBTT experiment using a few seconds' exposure of a piezoelectric quartz crystal which has been prepared for use as a frequency standard in a precision oscillator. Figure 11 shows a magnified view of one of the images on the previous figure. Since this beam line is instrumented with an image intensifier tube and TV camera, one can watch specific images change in real time as a result of imposed conditions (eg. temperature change, strain, etc.), while recording the TV output on a VCR for subsequent image analysis. The synchrotron has a particular utility for real-time white beam transmission topography when the crystals of interest have too high a mass absorption to be examined with conventional tube sources. One such case the authors have successfully examined has been single crystals of zinc cadmium telluride.

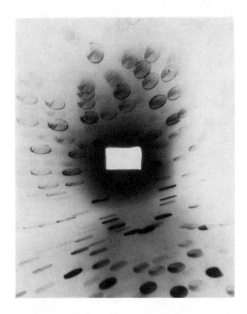

Figure 10 - White beam transmission topograph (WBTT)
of quartz cut for use in a precision oscillator.

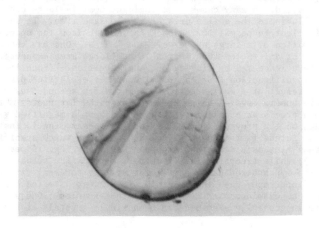

Figure 11 - Magnified image of one of the
Laue images in the preceding Figure.

Flash X-Ray Sources

In the following section, we will return to considering conventional Laue diffraction images, but with a different class of x-ray source. Figure 12 is a schematic of an image intensifier arrangement coupled to a flash x-ray (FXR) or field emission x-ray generator, also known as a pulsed x-ray generator (7).

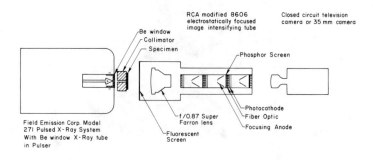

Figure 12 - Schematic of an image intensifier
arrangement coupled to a flash x-ray source.

This type of generator may have x-ray tube currents of the order of thousands of amperes as opposed to the tens of milliamps in conventional tubes. However, the burst of emitted radiation only lasts a few tens to perhaps a hundred nanoseconds. The result is extremely high x-ray photon flux over a very short span of time. Figure 13 shows a schematic of an experiment to record a diffraction pattern from the jet created by an explosively loaded shaped charge liner (8). The jet is formed by the collapse of a conical metal liner as it is deformed at a velocity in excess of the velocity of sound in the metal by the advance of a detonation wavefront created by an explosive charge packed behind the liner. The kinetic energy of such jets can be extremely high, so analytical models usually treated the jet as being totally liquid. But it was also known that metallurgical parameters (such as the initial crystallographic orientation in the liner before detonation) influenced the performance at the target. Figure 14(a) is a radiograph of such a jet, and Fig. 14(b) shows the corresponding diffraction pattern obtained from this jet while it is in flight. It shows that the jet is not a liquid.

Figure 15 shows the laboratory setup for an experiment described by Rabinovich and Green (9). Single crystals of ferroelectric materials usually consist of a number of oppositely polarized domains. A sufficiently high electric field, applied along the polar axis of the crystal, causes a polarization of the whole crystal in the direction of the field. When the direction of the applied field is reversed, a switching of the domains occurs, usually accompanied by a structural change. In this work, single crystals of gadolinium molybdate were chosen for dynamic studies of domain switching. A field of up to 30 kV per cm was applied and switched very quickly. Diffraction shots were taken at delay times ranging from a tenth of a millisecond after switching up to 2 milliseconds after switching. The successive images in Fig. 16 show the time development of the domain switching process as indicated by the real time diffraction images. The photographs are enlargements of the region of the x-ray diffraction pattern where the (560) plane weakly diffracts before polarity switching. The gradual build up of the strong (650) reflection is clearly evident.

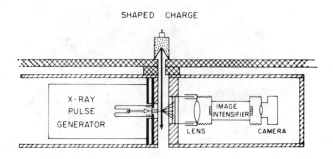

Figure 13 – Schematic of an experiment to record a diffraction
pattern from a shaped charge liner jet in real time.

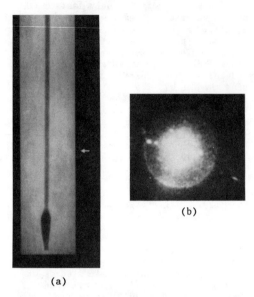

(a)

(b)

Figure 14 - (a) Flash x-ray radiograph of aluminum shaped charge liner jet. The white arrow indicates the position along the jet where the x-ray diffraction photograph was taken. (b) Transmission flash x-ray diffraction pattern of the aluminum jet.

Figure 15 - Laboratory arrangement for flash x-ray measurements of phase transformations.

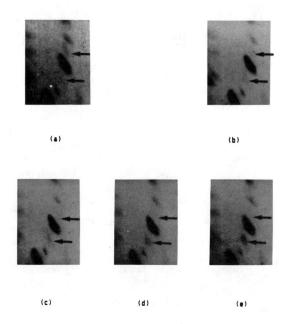

(a) (b)

(c) (d) (e)

Figure 16 - Time development of domain switching in gadolinium
molybdate crystal: (a) 0.1 msec, (b) 0.5 msec,
(c) 0.9 msec, (d) 1.5 msec, (e) 2 msec.

Computerized Tomography Using Flash X-Ray Sources

Although not an example of x-ray diffraction imaging, Fig. 17 shows a
design for what may be the ultimate in real time x-ray radiographic imaging,
and perhaps it shows a future possibility for diffraction imaging (10). It is
a schematic of the proposed layout for a computerized tomographic imaging
device using flash x-rays which would allow four orders of magnitude
improvement in temporal resolution over currently available systems. It would
use 21 simultaneously triggered flash x-ray sources and 21 detector arrays.
The design calculations predict a density resolution of several percent for a
1 cm region, lateral spatial projection resolution of 2 mm on a 200 mm object,
and a temporal resolution of one microsecond.

Summary

The generic configurations for monitoring real time x-ray diffraction
have been discussed. An example of the implementation of an indirect system
was shown which is able to document the degree of crystallization of rapidly
solidified alloys within a millisecond of solidification. Several
applications of real time imaging applied to x-ray topography were given,
including the use of both direct and indirect configurations. The use of
flash x-ray sources in conjunction with real time imaging was illustrated both
for recording the diffraction image from a shaped charge liner jet and for
recording rapid phase transformations. Finally, a proposed design for
computerized tomography using flash x-rays was described.

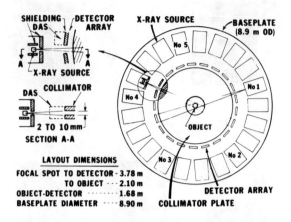

Figure 17 - A proposed design for computerized
tomography using flash x-ray systems.

Acknowledgements

    The synchrotron work described here was performed on the Synchrotron
Topography Project Beamline X-19C which is supported by the U. S. Department
of Energy under Grant No. DE-FG-02-84ER45098. The authors would like to thank
Mr. Wm. P. Hanson (of the authors' laboratory) for the use of the topograph
shown in Fig. 10.

References

1. R. E. Green, Jr., "Direct Display of X-Ray Topographic Images" in Advances
    in X-Ray Analysis, H. F. McMurdie, et al., eds., Plenum, N.Y.,
    20, (1977), 221-235.

2. R. G. Rosemeier and R. E. Green, Jr., "A New Miniature Microchannel Plate
    X-Ray Detector for Synchrotron Radiation", Nuclear Instruments and
    Methods, 195, (1982), 299-301.

3. A. R. Lang and K. Reifsnider, "Rapid X-Ray Diffraction Topography Using a
    High-Gain Image Intensifier", Appl. Phys. Lttrs., 15(8), (1969), 258-260.

4. J. Chikawa and I Fujimoto, "Video Display Technique for X-Ray Diffraction
    Topography". Nippon Hoso Kyokai (Japan Broadcasting Corporation),
    Technical Research Laboratories, Technical Monograph No. 33, March (1974).

5. J. Chikawa, I Fujimoto, S. Endo, and K. Mase, "X-Ray Television Topography
    for Quick Inspection of Si Crystals", in H. R. Huff and R. R. Burgess,
    eds., Semiconductor Silicon 1973, Electrochemical Society,
    Princeton, (1973), 448.

6. W. J. Boettinger, H. E. Burdette, M. Kuriyama, and R. E. Green, Jr.
    "Asymmetric Crystal Topographic Camera", Rev. Sci. Instr., 47(8), (1976),
    906-911.

7. J. A. Dantzig and R. E. Green, Jr., "Flash X-Ray Diffraction Systems", in
    <u>Advances in X-Ray Analysis</u>, L. S. Birks, C. S. Barrett, J. B.
    Newkirk, and C. O. Rudd, eds., Plenum, 16, (1973), 229-241.

8. Robert E. Green, Jr., "First X-Ray Diffraction Photograph of a Shaped
    Charge Jet", <u>Rev. Sci. Instrum.</u>, 46, (1975), 1257-1261.

9. R. E. Green, Jr. and D. Rabinovich, "Exploration of Structural Phase
    Transformationss by Flash X-Ray Diffraction", <u>1984 Flash Radiography
    Symposium</u>, E. A. Webster, Jr. and A. M. Kennedy, eds., The American
    Society for Nondestructive Testing, Columbus, Ohio, (1985), 157-170.

10. C. K. Zoltani and K. J. White, "Flash X-Ray Computed Tomography Facility
    for Microsecond Events", <u>Rev. Sci. Instrum.</u>, 57(4), (1986), 602-611.

USE OF NEUTRON POLE FIGURES TO CALIBRATE ULTRASONIC TECHNIQUES
FOR ON-LINE TEXTURE CONTROL OF ALUMINUM PLATES

R. C. Reno, A. V. Clark, G. V. Blessing, R. J. Fields
National Bureau of Standards

A. Govada
Aluminum Company of America

R. B. Thompson
Ames Laboratory, Iowa State University

P. P. Del Santo and R. B. Mignogna
Naval Research Laboratory

J. F. Smith
Iowa State University

Abstract

We have used neutron pole figures to determine the crystallographic texture
in rolled aluminum plates that were previously characterized with ultrasonic
techniques (1). The orientation distribution function coefficients (ODC's)
determined by neutron diffraction are in excellent agreement with those
derived from ultrasound measurements, thus confirming the efficacy of
ultrasound as an on-line monitoring technique.

Introduction

In crystalline materials with anisotropic elastic constants, the propagation of ultrasound is strongly dependent upon crystallographic texture. This dependence may provide investigators with a relatively rapid and economical method for monitoring texture in processed materials such as rolled plate (2-4). We have made use of ultrasound to characterize texture in rolled aluminum plates that are to be used in the fabrication of cans (1). Ultimately, it would be desirable to incorporate ultrasound-based texture monitors in the manufacturing process so that texture can be automatically controlled.

Although ultrasound propagation is influenced by texture, it is also affected by impurities, grain boundaries, and other inhomogeneities. It is therefore desirable to compare ultrasound texture measurements with more direct crystallographic measurements.

Neutron diffraction is an excellent method for studying texture in bulk samples directly. Neutrons penetrate deeply into materials, thus sampling the overall texture of specimens having volumes in excess of several cubic centimeters. The extremely high penetration in aluminum permits one to generate a complete pole figure without having to switch from reflection to transmission modes. Analysis of the pole figure results in orientation distribution function coefficients (ODC's) that can be compared to those measured with ultrasound.

Theory

In this paper, we will utilize the notation and methods developed by Roe (5) and Allen (6). We begin by defining an orientation distribution function (ODF), $w(\xi, \psi, \phi)$. Angles ξ, ψ, and ϕ are Euler angles which relate crystallite orientation with respect to bulk sample axes. In the case of aluminum plate, crystallographic axes are defined along the three cube directions of the aluminum unit cell and the sample axes are defined along the rolling, normal and transverse directions of the plate. The function $w(\xi, \psi, \phi)$ gives the proportion of crystallites whose orientation is within $\Delta\xi$, $\Delta\psi$, and $\Delta\phi$ of the specified Euler angles.

The orientation distribution function can be expanded in terms of generalized spherical harmonics, as given by Roe (5):

$$w(\xi, \psi, \phi) = \sum_{\ell=0}^{\infty} \sum_{m=-\ell}^{\ell} \sum_{n=-\ell}^{\ell} W_{\ell mn} Z_{\ell mn}(\cos \xi) \, e^{-im\psi} e^{-in\phi} \qquad (1)$$

where the $W_{\ell mn}$'s are orientation distribution function coefficients (ODC's) which quantitatively describe the crystallographic texture of a sample. Values of the ODC's are determined by measuring pole figures, $q_i(\zeta, \eta)$ and fitting the data to an expansion in spherical harmonics:

$$q_i(\zeta, \eta) = \sum_{\ell=0}^{\infty} \sum_{m=-\ell}^{\ell} Q_{\ell m}^{(i)} P_\ell^m(\cos \zeta) e^{-im\eta} \qquad (2)$$

Here the subscript i denotes a choice of Miller indices corresponding to a particular diffraction condition. ζ and η are polar and azimuthal angles describing the orientation of the sample (i.e. the plate normal) with respect

78

to the scattering vector.

The $Q_{\ell m}$'s obtained from a fit to the measured pole figure can then be used to determine the orientation distribution function coefficients ($W_{\ell mn}$'s) through the following relation:

$$Q_{\ell m}^{(i)} = (2\pi) \left(\frac{2}{2\ell + 1} \right)^{1/2} \sum_{n=-\ell}^{\ell} W_{\ell mn} \, P_\ell^n \, (\cos \Theta) e^{in\Phi} \qquad (3)$$

The angles Θ and Φ are polar and azimuthal angles describing the orientation of the reciprocal lattice vector (hkℓ) with respect to the unit cell crystallographic axes.

It is also possible to use ultrasonic measurements to determine the coefficients W_{400}, W_{420}, and W_{440} in hot-rolled aluminum plates. Consider the case of a rolled plate of polycrystalline material, and let the single crystals have cubic symmetry. On the macroscale, the plate can be assumed to have orthorhombic symmetry. The relations between the (cubic) single crystal constants and the nine orthorhombic second-order moduli have been derived by Sayers (2), using a Voigt averaging scheme. For the symmetries displayed by this system, the lowest order ODC's which are non-zero are the W_{4m0}, with m = 0, 2, 4; these are the only ODC's which affect the orthorhombic second-order moduli.

Consequently, measurements of sound velocity can be used to obtain the W_{4m0}. The theoretical relationship for velocities of bulk waves, guided-waves (plate-modes), and surface waves are given in Refs 1-4. For Lamb-waves and surface waves, W_{420} and W_{440} can be obtained by measuring the angular variation of velocity, i.e., by propagating ultrasonic waves at various angles to the rolling directions. The ODC W_{420} can also be obtained from the acoustic birefringence, which is the difference in velocities of orthogonally polarized shear-waves propagating through the plate thickness. The ODC W_{440} can also be obtained from angular variation of the lowest-order shear-horizontal plate mode. All of these techniques were employed in Ref. 1.

We should note that since W_{420} and W_{440} describe anisotropic properties of a plate with respect to plate normal, they can be measured with relative velocity measurements. Isotropic texture described by W_{400} requires absolute velocity measurements. For moderate textures such as in rolled aluminum alloy plates, it is much easier to obtain the required accuracy for ODC measurement using relative velocity measurements.

Furthermore, for application to on-line texture monitoring of rolled aluminum alloy product, we seek to measure only the minimum number of ODC's required. An example is the production of rolled sheet which will be used for can manufacture. Here the texture that is desired is one which leads to minimum wastage when pieces of the sheet are deep-drawn to our shape. Our ultimate goal is a correlation between the minimum number of W_{4m0}'s and a measure of plastic anisotropy during deep-drawing. For more details, the reader should consult Ref. 1.

Experimental Details and Data Analysis

Neutron pole figures were taken on two samples of hot-rolled aluminum plate, each having a different exit temperature. Complete pole figures were generated by placing the samples (0.5 inch diameter x 0.25 inch thick) in a beam of neutrons having a wavelength of 0.127 nm , selecting the appropriate detector orientation to observe diffraction from crystal planes defined by Miller indices (hkℓ), and measuring beam transmission over a range of sample orientations which span the entire hemisphere above the plane of the plate. Experiments were done at the NBS Reactor and data were converted to pole figures using programs written by C. S. Choi at NBS. Samples of pole figures taken at two different crystallographic orientations are shown in Figure 1.

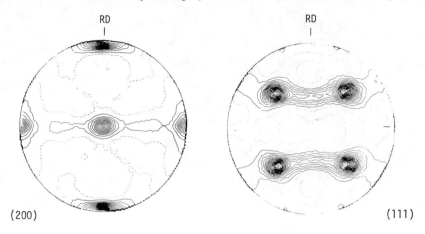

Figure 1 - Neutron pole figures of a rolled aluminum plate.
Two different diffraction conditions are shown.

Quantitative analysis of the pole figure data was accomplished with a program written by one of us (RCR) which inverts equation 2 and uses measured pole figure data to compute the $Q_{\ell m}$'s up to a maximum of $\ell=10$. The choice of maximum ℓ was dictated by computational considerations and the fact that ultrasound measurements only sense the $\ell=4$ coefficients. In order to see how well the original pole figure structure is reproduced with a series that is truncated at $\ell=10$, we have written a program that reconstructs a pole figure with specific $Q_{\ell m}$'s as inputs. Figure 2 shows reconstructed pole figures corresponding to the two measured pole figures shown in Figure 1.

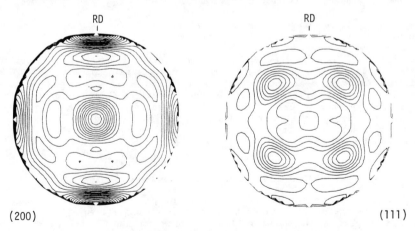

Figure 2 - Reconstruction of pole figures shown in Figure 1.
All components up through $\ell=10$ are included
in the reconstruction.

The reconstructions show the salient features of the measured pole figures, but obviously lack the high spatial frequencies that would have been present without truncation. The truncation does not, however, affect the values of the coefficients reported herein.

The inversion procedure provides us with coefficients (Q_{40}, Q_{42}, and Q_{44}) that are necessary to deduce the orientation distribution function coefficients that we require. For the case of aluminum plates, we can use crystallographic and sample symmetries to simplify equation 3 and obtain the desired ODC's. The simplified relations for the diffraction conditions used are given below.

For (111) pole figures: $W_{4m0} = - (3/4\pi) \, Q_{4m} = - 0.2387 \, Q_{4m}$

For (200) pole figures: $W_{4m0} = (1/2\pi) \, Q_{4m} = 0.1592 \, Q_{4m}$

For (220) pole figures: $W_{4m0} = - (2/\pi) \, Q_{4m} = - 0.6366 \, Q_{4m}$

For (311) pole figures: $W_{4m0} = 0.7407 \, Q_{4m}$

For (331) pole figures: $W_{4m0} = - 0.4288 \, Q_{4m}$

The ultrasonic measurements were made on a set of three hot-rolled aluminum plates (1). The set included the two samples tested with neutron diffraction. A variety of ultrasound techniques were used, as mentioned in the Theory section and they are summarized in Table I below.

Table I
Ultrasonic Measurement Techniques

ODC	Technique	Transducer	Configuration
W_{420}	Acoustic Birefringence	EMAT Piezoelectric	Pulse-echo
W_{420}	Lamb-wave	EMAT	Pitch-catch
W_{420}	Rayleigh-wave	Piezoelectric	Pitch-catch
W_{440}	Lamb-wave	EMAT	Pitch-catch
W_{440}	Shear-horizontal	EMAT	Pitch-catch
W_{440}	Rayleigh-wave	Piezoelectric	Pitch-catch

In the table, the acronym EMAT means electromagnetic-acoustic transducer, a device which generates ultrasound without use of an acoustic couplant. In pulse-echo configuration, the same transducer transmits and receives ultrasound; in pitch-catch, separate transmitting and receiving transducers (separated by a rigid spacer) are used.

Results

For each sample of aluminum plate, we generated five neutron pole figures and deduced the ODC's for $\ell=4$ using the method described in the above section. Table II shows the results from pole figure measurements, along with the values previously reported using ultrasound measurements. The agreement between the two methods is good. A comparison of Figures 1 and 2 show that reconstructions with components up through $\ell=10$ reproduce the salient features of the experimental data. Obviously, it is not possible to reproduce the data exactly unless a much larger range of ℓ's is employed. For this study, our concern is with the $\ell=4$ components only.

Conclusions

The good agreement between ultrasound and neutron pole figure data indicates that the ultrasonic measurements are indeed measuring aluminum plate texture, and are not being adversely affected by scattering from inhomogeneities. This further strengthens our previous conclusion (1) that ultrasound may be a viable method for monitoring texture in aluminum plate.

Table II
Comparison of Orientation Distribution Function Coefficients
Measured with Neutrons and Ultrasound

Method	W_{400}	W_{420}	W_{440}
Sample A (Exit Temperature 331 degrees Celsius)			
Neutron Pole Figures (this work)	+ 0.0096 ± 0.0024	- 0.0034 ± 0.0005	+ 0.0067 ± 0.0018
Ultrasound (Ref 1)	-----	- 0.0029 ± 0.0002	+ 0.0045 ± 0.0010
Sample B (Exit Temperature 357 degrees Celsius)			
Neutron Pole Figures (this work)	+ 0.0091 ± 0.0027	- 0.0032 ± 0.0008	+ 0.0043 ± 0.0012
Ultrasound (Ref 1)	-----	- 0.0025 ± 0.0001	+ 0.0028 ± 0.0010

Note: Uncertainties in pole figure data reflect spread in values deduced from five different pole figures; uncertainties in ultrasound data reflect variations in values measured by several techniques in several laboratories. Each measurement was given equal weight.

Acknowledgements

We wish to thank Dr. C. S. Choi (ARRADCOM/NBS) for his generous assistance during our early pole figure measurements.

The work reported here was sponsored at NBS and Ames Laboratory by the NBS Office of Nondestructive Evaluation, and at the Naval Research Laboratory by the Office of Naval Research.

References

1. A. V. Clark Jr., A. Govada, R. B. Thompson, J. F. Smith, G. V. Blessing, P. P. Delsanto and R. B. Mignogna, "The Use of Ultrasonics for Texture Monitoring in Aluminum Alloys", Review of Progress in Quantitative Nondestructive Evaluation, ed. by D. O. Thompson and D. E. Chimenti (1986), in press.

2. C. M. Sayers, "Ultrasonic velocities in anisotropic polycrystalline aggregates", J. Phys. D: Appl. Phys., 15 (1982) 2157-2167.

3. R.B. Thompson, J.F. Smith, and S.S. Lee, "Inference of Stress and Texture from the Angular Dependence of Ultrasonic Plate Mode Velocities", in NDE of Microstructure for Process Control, H.N.G. Wadley, ed., ASM, Metals Park, OH, 73 (1985).

4. P.P. Del Santo, R.B. Mignogna, and A.V. Clark, "Ultrasonic Texture Analysis for Polycrystalline Aggregates of Cubic Materials Displaying Orthotropic Symmetry," to be published in Proceedings of 2nd International Conference on Nondestructive Characteristics of Materials, Montreal, 1986.

5. Ryong-Joon Roe, "Inversion of Pole Figures Having Cubic Crystal Symmetry", Journal of Applied Physics, 37 (1966) 2069-2072.

6. A. J. Allen, M. T. Hutchings, C. M. Sayers, D. R. Allen and R. L. Smith, "Use of neutron diffraction texture measurements to establish a model for calculation of ultrasonic velocities in highly oriented austenitic weld material", J. Appl. Phys., 54 (1983) 555-560.

NONCONTACT ULTRASONIC SENSORS

FOR HIGH TEMPERATURE PROCESS CONTROL

G.A. Alers* and H.N.G. Wadley**

*Magnasonics, Inc.
215 Sierra Drive, SE
Albuquerque, New Mexico 98108

**National Bureau of Standards
Gaithersburg, Maryland 20899

Abstract

Ultrasound has been used to infer much about the internal
condition of materials (temperature, grain size, texture,
porosity, defects, residual stress etc.). To date, this
determination has been confined to studies under laboratory
conditions. If ultrasonic techniques could be developed that
could survive the high temperature aggressive environment of a
materials processing facility, the possibility exists of
developing a family of process control sensors to directly
probe microstructure and process variables. This paper reports
on the results of a joint AISI/NBS/Magnasonics program to
investigate the feasibility of using a Q-switched laser as a
source of ultrasound and an Electromagnetic Acoustic Transducer
(EMAT) for its detection on hot steel bodies. A variety of
permanent magnet EMAT designs have been evaluated. Direct
transmission experiments on both plain carbon and austenitic
stainless steels heated to as high as 980°C (1800°F) confirmed
for several of the EMATS, adequate signal-to-noise ratios and
bandwidths for sensor purposes.

Introduction

In 1982, the National Bureau of Standards and the American Iron and Steel Institute held a workshop to encourage the development of new sensor concepts that would increase productivity in the steel industry (1). Very high on the list of sensor needs presented by the AISI was a transducer that would allow ultrasonic tests to be carried out on thick billets in the temperature range of 800°C (1470°F) to 1400°C (2550°F). Such a device would not only allow internal defects such as pipe and porosity to be detected but it could also be used to infer the internal temperature distribution through the temperature dependence of ultrasonic velocity.

Previous attacks on the problem of developing such an ultrasonic transducer had shown that the requirement of intimate contact between conventional transducer devices and the hot steel rendered the final inspection system mechanically cumbersome and slow for the scanning of large areas. Thus, it was natural to consider noncontact types of transducers that could inherently avoid the problems associated with high temperature coupling. Two couplant-free sensors were seriously considered. One was based on optical techniques wherein a high power laser beam would interact with the surface to excite an ultrasonic pulse and interferometric or holographic procedures would be used to detect the pulse after it had passed through the hot steel. The second approach was to use electromagnetic coupling across a small air gap to both excite and detect ultrasonic tone burst types of acoustic signals. Although the former method clearly avoids physical contact with the steel and excellent generation of intense ultrasonic pulses could be demonstrated, optical techniques for detecting the pulses under steel mill conditions were considered premature and too costly to develop. The electromagnetic method was considered ready for development and had potential to withstand elevated temperatures. However, its electrical and magnetic characteristics had to be designed to overcome the lack of sensitivity inherent in the induction process, especially for the low electrical conductivities associated with very hot steel. A potentially optimum solution was to exploit the strong signal generating capacity of the laser source with the noncontact receiving capability of an EMAT to achieve a system of acceptable signal:noise ratio for sensor purposes.

This paper describes the results of a program aimed at demonstrating the feasibility of using a Q-switched laser as the ultrasonic transmitter and permanent magnet electromagnetic acoustic transducers (EMATs) as receivers. The results show that the combined sensor approach appears to be quite practical and very good signal-to-noise ratios were observed both on mild steel and stainless steel heated to 980°C (1800°F). The laser pulse contained only 175 millijoules of energy and the EMAT was a small, easily scanned probe designed around an array of small permanent magnets.

The Laser Transmitter

Excitation of ultrasonic waves by the absorption of an intense laser pulse is now a relatively well understood process (2). Research in England has shown that the generation process proceeds by two mechanisms (3). At low optical power, the metal surface is heated by the incident light radiation and ensuing thermal expansion causes a mechanical distortion to launch the acoustic disturbance. At higher powers, this thermoelastic mechanism still operates but now the surface temperature exceeds the vaporization temperature of the metal and the ejection of metal atoms from the surface imparts a momentum to the surface that generates an additional signal. This latter process is referred to as the ablation mechanism and is capable of generating ultrasonic pulses whose amplitudes substantially exceed those from traditional piezoelectric generators.

By using the theory of heat conduction, the thermoelastic mechanism can be modeled mathematically and the temporal surface displacement waveform of the ultrasonic signal can be predicted (3,4,5). Unfortunately, the ablative mechanism cannot be so completely described because the mass and velocity of the vaporized material is not well defined as a function of time. However, by making some reasonable assumptions, a temporal waveform with arbitrary amplitude can be predicted and agrees with experiment (2). In practice, it is not possible to produce pure ablation without a thermoelastic contribution so for most practical applications where large acoustic signal amplitudes are desired, the waveforms that are observed result from a combination of both mechanisms.

For the experiments described in this paper, a Q-switched Nd:YAG laser was used. It generated a 25 ns duration 1.01 μm wavelength pulse whose energy could be varied up to 800 millijoules. All of the experimental data reported here were obtained using a 175 millijoule per pulse energy and a 1 mm beam diameter.

The EMAT Receiver

An electromagnetic acoustic transducer or EMAT (6) detects ultrasonic waves in metals by an electromagnetic induction process across an air gap. Therefore, it is technically a noncontact transducer that operates without any couplant medium. It consists of a coil of wire held close to the metal surface plus a large magnet to supply a magnetic field that is used to convert surface motion into surface eddy currents which in turn couple into the coil by electromagnetic induction. For the present application to hot steel products, the air gap acts as a thermal barrier to reduce heat transfer into the EMAT structure and it allows the device to be scanned easily. In order to have high sensitivity, the magnetic field at the coil should be as large as possible. Therefore, it is the production of this field that dominates the design of the receiver device.

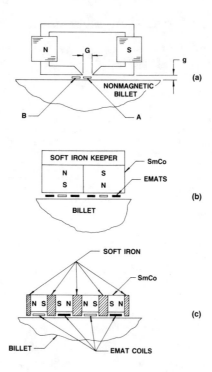

Figure 1 - Magnet configurations needed to operate EMATS:
(a) electromagnet for detecting longitudinal waves (A) or
shear waves (B); (b) permanent magnet arrangement for
detecting shear waves; (c) permanent magnet design for
detection of longitudinal waves.

Magnet Design

The most direct approach to supplying a magnetic field to
an EMAT coil is to use a large electromagnet with pole pieces
designed to focus the magnetic flux to the coil as shown in
Figure 1(a). Here, the EMAT coil can be located where the
field is in either a normal or tangential direction. Position
A, in a tangential field, is suitable for reception of
longitudinal waves while Position B, in a normal field, is
suitable for shear waves. Unfortunately, such an electromagnet
is quite massive and difficult to move in a scanning apparatus
even though it can generate magnetic fields in the 10 kilo-
oersteds range at an EMAT coil with a 1/2-inch diameter.
Figures 1(b) and 1(c) show alternate magnet configurations
based on permanent magnets located very close to the EMAT coil
in order to maximize the field at the coil wires. Both of
these structures were assembled out of 1/4-inch thick by 1/2-
inch slabs of samarium cobalt and fields in the range of 3 to 4
kilo-oersteds were measured at the EMAT coil positions. These
permanent magnet structures were quite light and maneuverable
and thus could be easily scanned over large areas. It turns
out that their reduction in efficiency caused by smaller

magnetic fields could be tolerated in exchange for this simplification in mechanical structure. No active cooling by water circulating past the magnets was used because it was found that the thermal inertia of the assembly itself was sufficient to permit contact with hot steel for periods of time greater than one minute without harming the performance of the EMAT probe.

Coil Design

The EMAT coils were wound in the shape of a series of flat slabs in order to fit under the magnets and minimize the air gap (as shown in Figure 2). Each slab was formed out of a row of 1/16-inch diameter thermocouple tubes held together with a high temperature adhesive. Each tube contained four small holes into which copper wires were threaded to form the EMAT coil itself. The direction of winding for each slab and the connections between slabs were determined by the particular EMAT desired as described below.

The permanent magnets shown in Figure 1(c) apply a tangential magnetic field to the surface of the part being inspected and this field changes direction in adjacent cells. Therefore, if the EMAT coil slabs that fill the spaces between the iron pole pieces are all wound with wires running in the same direction, the result is an EMAT coil with a periodicity in its field equal to the spacing between pole pieces (a distance of 1/4 inch for the case described in this paper). Such a structure is sensitive to sound waves that approach the plane of the EMAT at an angle θ relative to the surface normal given by the equation:

$$\sin \theta = \frac{V_S}{2Df} = \frac{V_L}{2Df} \qquad (1)$$

where V_S and V_L are the shear and longitudinal ultrasonic wave velocities in steel respectively, D is the half period of the magnetic field (1/4 inch in this case) and f is the frequency of operation.

If the coil wires in adjacent spaces are wound in opposite directions, the sense of the winding and the magnetic field change direction simultaneously across the face of the probe and there is no periodicity. Such an arrangement has optimum sensitivity for longitudinal waves that approach the face of the transducer along the normal to the surface. Thus, it is a transducer very well suited to detecting the longitudinal wave generated by a laser ablation source on the surface directly opposite to the sensor. For EMATs to be sensitive to shear waves, the normal field configuration shown in Figure 1(b) must be used. Here, adjacent coil slabs were wound in an opposite direction when under the same magnet pole as shown in Figure 1(b) in order to yield an EMAT sensitive to waves at the angle θ given in Equation 1.

Choice of Frequency

One important problem to be overcome when designing an EMAT for operation on hot steel is to be sure that the electrical resistivity of the hot steel is not so high that the electromagnetic skin depth become comparable with the wavelength of the ultrasonic waves involved (8). For ordinary metals with resistivities near 10 µohm-cm, this criterion is easily satisfied for frequencies less than 10 MHz. However, hot steel has a resistivity between 100 and 200 µohm-cm so the frequency should be kept less than 1 MHz for longitudinal waves and 0.3 MHz for shear waves. A second problem is related to the attenuation of the sound waves in the hot metal. Papadakis (9) has shown that the attenuation of longitudinal waves in the low megahertz range is less than for shear waves at the same frequency so sensitivity is probably better if longitudinal waves are used. If the transit time is to be measured, the better time resolution available through the use of higher frequency waves would also indicate the use of longitudinal waves.

Room Temperature Tests

Specimen Configuration

Austenitic stainless steel at room temperature does not exhibit ferromagnetism, has a high electrical resistivity and has an acoustic impedance and is thus similar to hot plain carbon steel. Thus, it represents a good simulation of carbon steel at a very high temperature. Figure 2 shows the geometrical configuration used to test various laser and EMAT concepts. For angular dependence studies, a half cylinder of 304 type stainless steel, 3 inches in radius and 3-inches high was mounted on an optical bench so that the laser beam would impinge on the curved surface along a radial direction as shown in Figure 2(b). In this configuration, the sound wave excited by the laser could propagate along the radius of the cylinder and approach the EMAT at an angle relative to the surface normal of the EMAT. By rotating the cylinder about its axis, the amplitude of the output of the EMAT and the arrival time of the acoustic energy could be measured as a function of angle. From the arrival time, it was possible to determine the type of wave detected by the EMAT.

Beam Angle Dependence

The two permanent magnet EMATs shown in Figures 1(b) and 1(c) with magnetic fields normal and tangential to the surface of the part were tested in the assembly shown in Figure 2(b). The EMAT coil slabs under the magnets were wound in a meander fashion with adjacent slabs carrying current in opposite directions as indicated by the open and closed boxes in Figure 1. In this way, tests could be performed with waves propagating at various angles to the surface normal by choosing different frequencies of operation as prescribed by Equation 1. Figure 3 shows the theoretically predicted angular dependence

of EMAT efficiency for both longitudinal and shear waves for
both directions of magnetic field (7). Also shown in this
figure are the predicted dependence of the acoustic energy
radiated by a pulsed laser source (3). Since there are two
mechanisms of launching elastic waves by the absorption of
optical energy, there are two distinctly different angular
distributions for the laser generated acoustic energy.
Comparison of the angular distribution functions show that at
low optical energy densities where the thermoelastic mechanism
dominates, the angular dependence is very similar to an EMAT

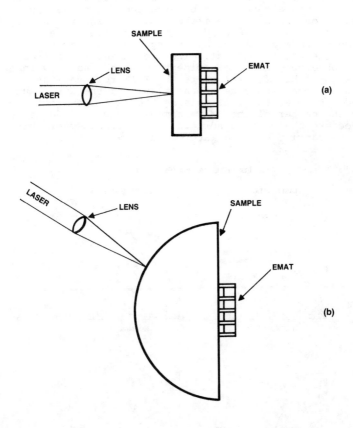

Figure 2 - Specimen configurations used to test the
efficiency of laser transmitters and EMAT receivers of
ultrasonic energy: (a) high temperature tests; (b)
ambient temperature tests.

operating in a normal magnetic field while for the high energy, density ablation mechanism, the angular distribution is like an EMAT operating in a tangential magnetic field.

In accord with the ablative mechanism of generating ultrasonic waves, very large ultrasonic signals were observed with the EMAT shown in Figure 1(c) with its coil slabs wound to detect longitudinal waves incident on the EMAT along the surface normal as shown in Figure 2(a). Another large acoustic signal was observed to be a shear wave propagating near 30 degrees to the surface normal and was detected by the EMAT with its magnetic field perpendicular to the surface as shown in Figure 1(b). Such a wave would be expected for either the thermoelastic or the ablative mechanism. One surprising result was the observation of a second large shear wave signal leaving the surface of optical impact at an angle of 60 degrees relative to the surface normal. These high amplitude shear waves were observed in the experiments with the half cylinder as well as in slabs of steel where the EMAT receiver was located on the opposite side of the slab from the laser but positioned such that the line joining the laser beam impact point and the center of the EMAT formed an angle of 60 degrees with respect to the surface normals. Further study of this wave will be carried on at a later time.

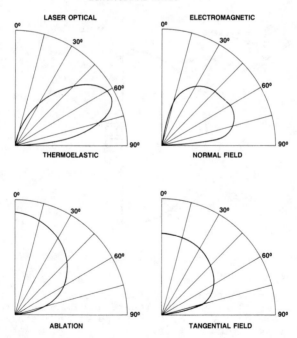

Figure 3(a) – Angular dependence of longitudinal waves expected to be excited by the two optical mechanisms and detected by EMATs with different orientations of the magnetic field relative to the sample surface.

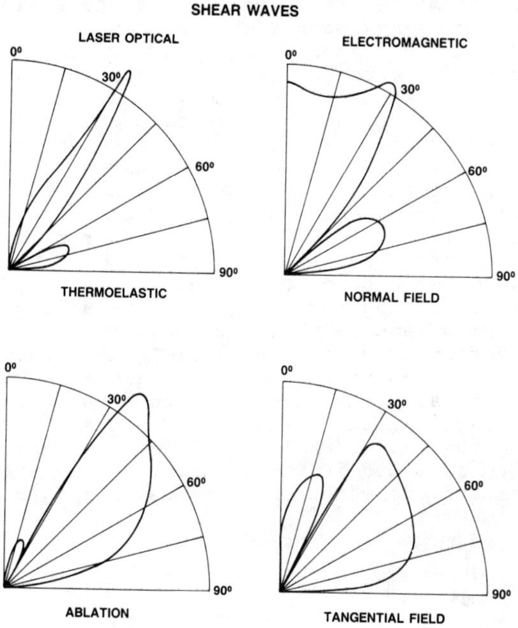

SHEAR WAVES

LASER OPTICAL   ELECTROMAGNETIC

THERMOELASTIC   NORMAL FIELD

ABLATION   TANGENTIAL FIELD

Figure 3(b) - Angular dependence of shear waves expected
to be generated by laser beam impact and detected by EMATs
with different magnetic field orientations relative to the
sample surface.

Air Gap Dependence

A very important performance characteristic of the
longitudinal wave EMAT shown in Figure 1(c) was the amount of
air gap or lift-off that could be tolerated between the sample
and the front face of the EMAT. Figure 4 shows how the output
signal from the EMAT decreased as a function of this separation
distance. This graph shows that the addition of up to 0.02
inches of thermally insulating material between the pole pieces
and the hot object would not cause too significant a reduction
in sensitivity but would probably make a dramatic improvement
in the ability of the structure to withstand exposure to high
temperatures. The fact that the drop in signal strength is
described by an exponential function is consistent with the
periodic structure of this form of magnet. The rate of drop
could be decreased to achieve higher sensitivity at large air
gaps by using thicker permanent magnets to increase the
separation distance between pole pieces.

Frequency Bandwidth

Another performance characteristic of the EMATs shown in Figure 1 is the frequency bandwidth that can be achieved with various coil designs and amplifier circuits. For EMATs made with meander coils to detect ultrasonic shear waves at an angle to the surface as shown in Figure 1(b), a narrow bandwidth is preferred in order to define the direction of wave propagation (as described by Equation 1) and to permit narrow band amplifiers to be used. For longitudinal waves propagating perpendicular to the surface, the EMAT shown in Figure 1(c) will accept all frequencies. Thus, broad bandwidth operation is not only possible but is very desirable for time-of-flight measurements. Figure 5 shows the waveforms observed when the longitudinal wave EMAT shown in Figure 1(c) was used to receive directly transmitted longitudinal waves generated by the pulsed laser. When narrow band filters were used in the amplifier stages, the waveform shown in Figure 5(a) was observed. Note that the noise received prior to the ultrasonic wave arrival is very small and the acoustic pulse is a tone burst containing many cycles of oscillation. Figure 5(b) shows the effect of simply removing the filter while retaining a capacitor across the EMAT coil to tune out its inductance. Here, the noise level is higher but the acoustic pulse is no longer a tone

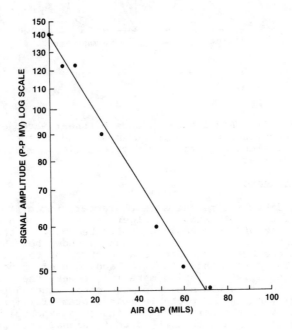

Figure 4 - Dependence of the sensitivity of the permanent magnet EMAT shown in Figure 1(c) upon the thickness of the air gap between the sample and the EMAT.

burst and its arrival time can now be accurately measured. The largest bandwidth was obtained by removing the tuning capacitor from the EMAT coil and the very sharp waveform shown in Figure 5(c) was observed. Note that the time scale has been expanded for this figure and that the noise level prior to the arrival of the acoustic wave is much higher than in the previous cases. Each printed waveform is the result of averaging eight individual waveforms generated by eight laser pulses.

Elevated Temperature Results

In order to demonstrate laser excitation and EMAT reception of ultrasonic waves in hot steel, several block-shaped samples ranging in weight from 15 to 30 pounds were soaked in a large furnace at 980°C (1800°F). These hot blocks were carried to the optical bench and, as quickly as possible, positioned such that the focal point of the laser beam was at the center of the front surface of the sample. The EMAT was then rested lightly against the back surface directly opposite the laser beam impact point as shown in Figure 2(a). Thermocouples were held against the sample surface and inserted into the EMAT structure at the location of the samarium cobalt permanent magnets in order to monitor the local temperatures while the steel cooled and the EMAT heated up.

Since it was desired to make time-of-flight measurements and because the attenuation of high frequency shear waves in hot steel is higher than longitudinal waves (9), only the comb magnet construction technique shown in Figure 1(c) was used in the high temperature tests. All the wires in each slab of the EMAT coil between the pole pieces conducted current in the same direction but this direction reversed in the adjacent pole piece gap where the magnetic field was also in the opposite direction. In this way, the entire face of the EMAT would respond in phase to a plane longitudinal wave striking the face along its normal direction. This method of assembling the permanent magnet structure allows the pole pieces to conduct heat directly to the samarium cobalt. Thus, the ability of the probe to operate on hot objects depends critically on the length of time it takes to heat the permanent magnets to a temperature at which their performance is jeopardized. The manufacturer of the magnets used in these EMATs set the upper temperature limit for this product at 350°C (660°F) for continuous operation. During the tests on hot blocks that were cooling from 1000°C (1830°F), a thermocouple on the samarium cobalt showed that it required over 1-1/2 minutes for the temperature of the magnets to exceed 250°C (480°F). This 90 second time interval was ample time for collecting all the necessary ultrasonic information on the samples. Therefore, it was not necessary to add water cooling apparatus to the EMAT probe.

Two samples were used for testing the response of the EMAT to elevated temperature operation. One sample was made of AISI 304 stainless steel and hence underwent no ferromagnetic

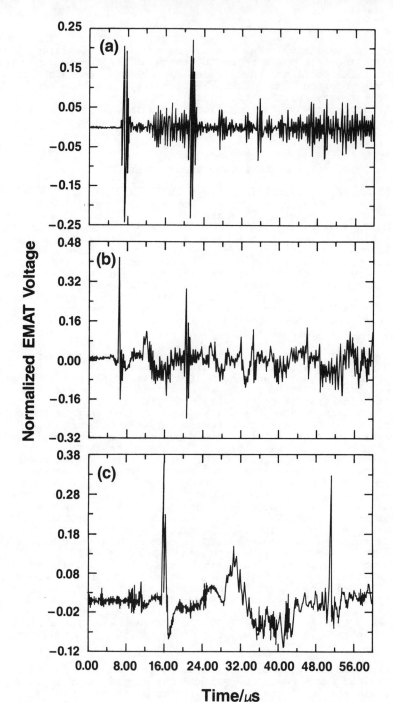

Figure 5 - Ultrasonic waveforms observed with different
electronic tuning circuits connected to the longitudinal
EMAT coil: (a) band pass filter combined with a tuning
capacitor across the EMAT coil; (b) only a tuning
capacitor at the EMAT; (c) no tuning.

transition as it cooled. This sample was a 4-inch diameter by
4-inch long cylinder in which the sound wave propagated along
the axis of the cylinder. The second sample was a 6-inch by 6-
inch square slab of AISI 1018 steel arranged so that the sound
wave propagated parallel to the 4-inch thickness dimension.
For this sample, a transition from the nonmagnetic to magnetic
state would be expected when it cooled below the Curie
temperature of 770°C (1418°F).

Figure 6 shows the waveforms observed on the 4-inch long,
stainless steel cylinder as its surface temperature fell from
752°C (1386°F) to 322°C (611°F). The directly transmitted
longitudinal wave signals and three reverberations can be
easily distinguished from the background noise. Most of this
noise is probably from ultrasonic signals reflected by the side
walls because the noise is at the EMAT frequency and arrives
after the direct longitudinal wave. Note that the time
interval immediately following the laser pulse and prior to the
arrival of the direct longitudinal wave signal is very quiet at
all temperatures. As the sample cooled, the acoustic noise
between the reverberation signals increased as would be
expected if there was a significant lowering of the attenuation
for ultrasonic waves as the temperature of the steel became
lower.

Figure 7 shows the waveforms observed on the 4-inch thick
slab of 1018 steel as the surface temperature dropped from
810°C (1490°F) through the Curie temperature to 545°C (1013°F).
At the highest temperature studied, the ultrasonic longitudinal
wave signals are very well defined and appear similar to the
signals observed on stainless steel. Below the Curie
temperature, the magnetic field from the pole pieces of the
comb-type EMAT would be expected to flow directly into the
steel in the immediate vicinity of each pole piece so that the
tangential field at the EMAT coil would become greatly reduced.
Therefore, the sensitivity to longitudinal waves should fall
and the directly transmitted longitudinal wave signal which
should arrive at about 20 μsec should no longer be the dominant
signal observed. Instead, a late arriving pair of signals were
observed and could be interpreted as shear waves that had
reached the EMAT by reflecting from the side walls of the
sample. The arrival times of about 50 and 70 μsec can be made
consistent with this hypothesis if the shear waves were
launched at an angle to the front surface normal and reached
the EMAT by multiple reflections from the side wall. Since
these shear waves have reflected from the side of the sample,
they must impinge on the EMAT at an angle so Equation 1 must be
satisfied and a periodic normal field must exist in the EMAT
structure. Such a field is actually present in the immediate
vicinity of the pole pieces of the comb-type EMAT when the
sample is ferromagnetic. An analysis of the geometrical
dimensions of the EMAT used in these experiments indicates that
a periodic magnetic field perpendicular to the surface is
present with a probable spacing of 0.17". This would make the
EMAT sensitive to shear waves that approach the face of the
probe at an angle of about 48 degrees relative to the surface

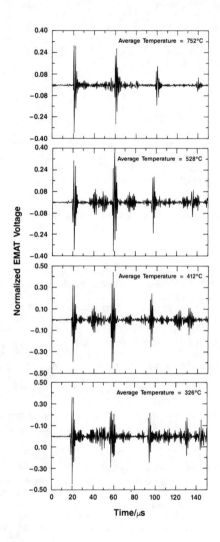

Figure 6 - Waveforms observed on a 4-inch thick stainless steel block at four temperatures.

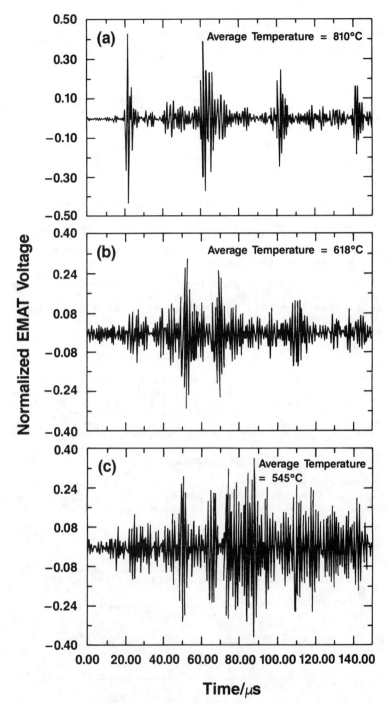

Figure 7 - Waveforms observed on a 4-inch thick block of
1018 ferritic steel as it cooled through the Curie
temperature.

normal.and is consistent with the sound wave paths reflected by the side walls of the sample.

Conclusions

1. A lightweight, easily scanned EMAT receiver probe was constructed out of heat resistant materials and samarium cobalt permanent magnets.

2. This probe was able to detect at ambient temperature, ultrasonic pulses generated by a 175 millijoule pulsed laser with a signal-to-noise ratio of 40:1 (32 dB) after propagating 4-inches.

3. Both a mild steel block and a stainless steel block were tested on a laboratory optical bench as they cooled from a furnace temperature of 980° (1800°F). Signal-to-noise ratios similar to those observed at room temperature were observed for longitudinal waves but not for shear waves when the samples were at their highest temperatures.

4. No degradation of performance was observed after the probe had been repeatedly held in contact with the hot steel samples for time intervals that ranged from 60 to 90 seconds for each contact even though no active water cooling was employed.

Acknowledgements

Floyd Mauer and Jeff Martinez deserve special recognition for their operation of the apparatus, the handling of hot steel slabs and their help with analysis of the data. Dr. Bernie Droney of Bethlehem Steel Corporation provided important insights into the phenomena that were observed and helped to focus research toward steel industry sensor needs.

References

1. "Process Control Sensors for the Steel Industry", A Workshop Sponsored by AISI, DARPA, and NBS at the National Bureau of Standards, Gaithersburg, MD, July 27-28, 1982.

2. H.N.G. Wadley, Journal of Metals, Vol. 38, No. 10, October (1986), 49-53.

3. C.B. Scruby, et al., Research Techniques in NDT, Vol. 5, ed. R.S. Sharpe, Acad. Press, Chapter 8, (1982), 281.

4. H.N.G. Wadley, et al., Review of Progress in QNDE, 3B, (1984) 683, Plenum Press, New York.

5. C.B. Scruby, Appl. Phys. Letters, 48 (2), 100, January (1986).

6. R.B. Thompson, Proc. 1973 Ultrasonics Symposium, IEEE Cat. No. 73, CHO 807-8SU.

7. C.F. Vasile and R.B. Thompson, <u>Ultrasonics Symposium Proc.</u>,
    1977, IEEE Cat. No. 77CH1264-ISU, 84 (1977).

8. R.E. Beissner, "EMATs – A Survey of the State of the Art",
    Nondestructive Testing Information Analysis Center,
    Southwest Research Institute, San Antonio, Texas, NTIAC-
    76-1, January (1976).

9. E.P. Papadakis, et al., <u>J. Acous. Soc. Am.</u>, 52, 855 (1972).

DIRECT MOLTEN METAL ANALYSIS BY TRANSIENT

SPECTROSCOPY OF LASER PRODUCED PLASMAS

Yong W. Kim

Department of Physics, Bldg 16
Lehigh University, Bethlehem, PA 18015

ABSTRACT

Real time analysis of molten metal for elemental composition is of great interest from the standpoint of process control in metals production, as well as dynamincal studies of solidification. The extremely harsh steel-making environment presents a potentially useful opportunity for establishing a new methodology of analysis with broad applicability. The basic concept under investigation is to create an optically-thick plasma plume off the molten metal surface by means of a laser pulse and measure the elemental abundances by transient spectroscopy. Experiments with both molten and solid metal samples are being carried out in the laboratory in conjunction with numerical simulation based on a numerical code which incorporates laser heating and evaporation of condensed-phase metallic species, multi-photon ionization, laser-plasma interaction and transport of particles, momenta and energy, including radiation transport and excitation transfer. Considerable attention has been given to the logistics of bringing together the laser source, spectroscopic instrumentation and fast electronics including computers in a survivable probe package. It appears at present that the measurement, data analysis, decision making and priming of the probe for the next round can be carried out in less than a minute.

INTRODUCTION

Steel making involves the basic steps of charging the furnace, melting the charge, and refining. Making molten metals and other alloys requires basically similar steps. During refining, it is critical that the operating parameters be adjusted and controlled so that the chemistry of the melt is within predetermined units of elemental concentration. The conventional method of taking a sample from an operating furnace and having it analyzed is labor intensive and time consuming, often taking up to 20% of the overall time required to refine a batch of metal. Furthermore, there are certain uncertainties introduced into the analysis due to partial elemental segregation which may result during the cooling of the sample.

The alternative approach, which is both rapid and information-rich, is to analyze the molten metal directly under in-situ, real-time conditions. Such an approach can provide a basis for bona fide real-time process control for metal production.

A successful technique for molten metal analysis must be applicable under the extreme high temperature conditions that exist inside a metal furnace. It must also be able to overcome a great number of variations that exist of the thermal and fluid-dynamic state of the molten metal and of the chemical properties of the slag or other oxide layer above. In addition, the technique cannot rely on any physical phenomena which depend sensitively on the physical properties of the molten metal such as shear viscosity, surface tension, elemental vapor pressure and sound speeds. Of course, any sensor elements employed in the technique must either be able to survive the bath temperature or be provided with cooling without risking the freeze-up of the slag or molten metal on them.

These requirements eliminate virtually all but the two following approaches: a) excitation and subsequent examination of the particulate and gaseous effluents from the molten metal bath, which are either naturally produced or artificially generated, and b) rapid generation, and atomic excitation, of a vapor plume from a slag-free molten metal surface by an intense laser pulse, followed by spectroscopic analysis of the emission spectrum. The first approach is advantageous in that measurement activities may take place outside a given furnace by transporting the effluents in a gas flow, thus allowing for extensive instrumentation. This approach is potentially well-suited for generating a great deal of process data.

It is, however, burdened with the need to determine under real-time conditions the extent of elemental contributions to the effluents by the slag in relationship to the molten metal. Our earlier investigation into the mechanisms of particulate production in steel furnaces has shown that the relative contributions depend strongly on the intensity of gas bubbling in the molten metal bath and the furnace temperature profile as well as the slag composition and the nature of the nucleation centers from which the final particulates have evolved.[1,2] The size and composition distribution of the particulates also depend on their residence time in the vapor-rich region of the furnace whose elemental abundances are determined by both the

104

elemental evaporation rates and the elemental abundances in the molten metal. While there are ways to clarify all of these ambiguities, it is clear that an extensive program of sorting out all of the competing processes must be carried out first before addressing the primary task of metallic composition measurements.

In this paper we describe an investigation primarily along the line of the second approach to rapid in-process determination of the elemental composition of the molten metal within a metal furnace. The concept is first to penetrate the slag layer by means of an ablatively protected sensor-probe and then to heat a small area of the exposed molten metal surface by means of a focused beam of a pulsed high-power laser. The objective of such heating is to produce a plasma plume having the same elemental composition of the molten metal. The emission spectrum from the plasma plume is analysed by means of time-resolved spectroscopy and the elemental abundances are determined for all elements of interest to a given metallurgical process.

In the following sections we describe the overall experimental arrangement, the physics of the plasma plume formation, radiation transport through the plasma, and the theoretical modeling by means of numerical simulation. A brief summary is given of the current status of the research at the end.

We note here that the present approach to elemental analysis of the molten metal addresses the two separate issues of interest in a significant way. First of all, use of the laser-produced plasma plume represents a fundamentally new methodology of elemental analysis with a number of unique features, not available in the conventional methods of inductively coupled plasmas or spark discharge plasmas. Secondly, in this approach the total analysis time is reduced to an unprecedented under-a-minute range for multi-element metallic alloys in a molten state. In addition, the extremely harsh steel-making environment as the normal operating condition for this methodology serves a useful purpose in that its successful implementation will pave a way to the widest possible range of applications.

LASER PRODUCED PLASMAS - EXPERIMENT AND MODELING

The laboratory investigation is centered around an experimental setup which, in broad categories, consists of a high-power Nd:glass laser, two different metallic targets of both solid and molten form, spectroscopic instruments, assortments of time-resolved detectors, diagnostic and control electronics, and laser beam and light handling optics.

The molten metal target is prepared inside a vacuum chamber, which houses an induction furnace, in order to keep its surface free of slag. The laser beam is focused onto the molten metal surface from above through a single lens and the optical emissions from the laser produced plasma are observed through several optical ports over the full spectral range from the ultraviolet to near intrared. During the entire period of a heat a flow of an inert gas is maintained through the vacuum chamber in a manner designed to control the diffusion of metal vapors.

In the case of a solid target, the laser beam is directed toward the target by means of a second set of optics. All other diagnostics are similarly redirected for the solid target experiment.

The laser is operated in a short pulse mode of approximately twenty nanosecond duration. The total energy delivered in a single pulse is as large as twenty Joules. The primary criterion for the operating condition of the laser is to provide the laser heating rate such that it is the single dominant rate-controlling element in the process of plasma formation from metallic targets. We will expand on this point further. The early part of the laser energy pulse is consumed in the rapid heating of the target metal surface so that vaporization of the condensed phase species can be initiated and sustained. In the cases where marginally intense laser pulses are delivered, conduction of the heat into the bulk competes significantly with the process of evaporation. A net result is then a modest rate of evaporation of the target metal with an attendant sensitivity to the elemental dependence of the rate of evaporation. In other words, the elemental composition of the vapor becomes shifted in favor of those more volatile species away from the composition in the condensed phase. This is an undesirable situation, and is avoided by choosing the laser pulses in such a way that the evaporation front keeps pace with the thermal conduction front in the bulk.

The primary thrust of the diagnostic measurements has been in determining the intensity of the emission spectrum from the laser-produced plasma as a detailed function of time. The reason for this emphasis has to do with the process of radiation transport within the plasma during the period of formation by laser heating and the subsequent period of afterglow when the plasma undergoes decay. When the plasma reaches its peak temperature in the neighborhood of 10^6 K, the electron density may be as high as 10^{23} electrons/cm^3 and the plasma core radiates as a blackbody radiator. One can understand the continuum radiation as resulting from a) the Stark broadening of the elemental emission lines due to thermalized interactions of radiating particles with electrons and ions, b) the emissions due to free-to-bound-state transitions and c) the Bremsstrahlung process involving free-to-free-state transitions.(3) In any event, the transport of the radiation from the plasma core is diffusive in space at such high densities. Furthermore, the radiation in one given narrow range of wavelength may be absorbed by atoms and ions by resonance transitions, leading to their internal excitation, and such excited state atoms and ions collide with electrons and other particles and transfer excitation to those in the form of the internal as well as kinetic energy. The net result is that the radiation transport takes place along the energy axis as well in the form of collision-mediated excitation transfer.

The spectroscopic measurement of elemental concentration in

the plasma generated from a metallic alloy is based on a stable cause-and-effect relationship between the emission intensity at the given wavelength of a resonance transition line belonging to an elemental atom and the number density of the atoms in the plasma. The radiation transport processes described above alter the relationship in a time-dependent manner because the emission intensity at a given wavelength may be expended in part due to the excitation transfer process and in another due to its transport impeded by the radiation diffusion process. The degree to which these processes of radiation transport manifest themselves in the detected intensity of the resonance emission depends on the temperature and electron density of the plasma and is therefore a strong function of time. It is quite clear then that any spectroscopic measurement of elemental concentration must first identify the regions of time, somewhere between the birth and the decay of the pulsed plasma plume, in which the emission intensity and the number density of the radiating particles are stably related.

A dramatic manifestation of the excitation transfer process is the line reversal phenomenon. The continuum or severely broadened line emission from the plasma core becomes absorbed in the cooler outer region, according to an absorption profile determined by the local temperature and the number densities of electrons and those absorbing atoms. The detected spectrum, in the vicinity of the resonance line of the atomic species under discussion, exhibits a feature of missing intensities around the line center. The depth of the missing intensity profile as measured with respect to the background can be related to the number density of the particular atomic species. The background intensity is determined from the intensities at the line wings.

Extensive laboratory measurements have shown that the laser-produced plasmas exhibit a strong continuous spectrum early in the evolution, peaking out in intensity shortly after the end of the laser pulse. This is followed by an extended afterglow regime where the emission spectrum rapidly loses its continuum character and attains the dominant line character. The afterglow regime lasts for about ten microseconds. The line reversal has been observed for most of the resonance lines of the elemental species commonly used in metallic alloys.

Of critical importance from the standpoint of the real-time in-situ measurement strategy which we are pursuing is the fact that the laser plasmas have been found to be much more robust than any electrically produced plasmas, which are commonly used for the conventional analysis schemes. The emission spectrum is intense enough for complete spectral determination from a single laser pulse and the emission lines of multiply ionized species are readily detected. Of course, presence of the strong continuum emissions demonstrates a very high plasma temperature which is normally inaccessible by conventional spark discharges. In essence, we have shown that under our laboratory conditions information-rich plasmas can be generated, representative of the elemental composition of the target metal of both the solid and molten state.

The steep rise to the state of the continuum emissions over the entire spectral range of interest, within the twenty nanosecond duration of the laser pulse, is particularly significant because it demonstrates that the desired laser heating rate has been attained. That is, the laser heating rate is sufficient to evaporate the target metal layers in pace with the thermal conduction front in the bulk of the metal.

Simulation by an independent numerical code(4,5) predicts such a connection when the laser intensity profile, the metal target properties and the properties of the gas in contact with the metal surface are defined as the initial condition of the code. Under these conditions the code also shows that the condensed phase metallic elements are heated to states above their critical points over the distances into the bulk, which are greater than the distances of sound propagation during the same time interval. This is entirely consistent with the laser heating requirement discussed earlier. The validity of the simulation code is verified by experimental confirmation of those measurable predictions from the code. One such prediction is that the solid target plasma attains a lower plasma temperature and density than that from the molten metal target due to the difference in the energy expenditure per unit mass by the latent heat of melting between the two. The experimental results indeed show such a difference in the peak continuum intensity over the entire spectral range of interest.

THE CURRENT STATUS OF THE PROBE DEVELOPMENT

The elements of interest for real-time in-situ analysis are Ca, Ti, Na, V, Mn, Fe, Mg, Cr, Cd, Al, Ni, B, Cu, Zn, C, Si, S, and P. The resonance lines of these elements are accessible in the uv to visible spectral range and therefore are candidates for simultaneous concentration analysis. The ultimate accuracy and the detection limit, subject to the typical range of concentration in common alloys, are presently evaluated using the laser-produced plasmas as the spectral source.

Detailed evaluation of the feasibility of carrying out the entire measurement activities and subsequent analysis for concentration of all elements of interest under one minute has been made. The measurement activities include ranging and firing of the laser at the moment when the molten metal surface is presented at the focal distance from the laser beam focusing lens, time-resolved detection of the plasma emission spectrum, transfer of the spectral data to a dedicated computer, and rapid analysis of the data to determine the elemental concentrations of the molten metal. Each of these activities has been laboratory tested from both the hardware and software standpoint. We have determined that integration of the above measurement activities can indeed facilitate rapid in-process analysis of molten metal under one minute.

Preparations are being made for field trials of this new methodology. A probe is presently under construction, which houses all measurement instrumentation and is protected from the extreme heat of the metal furnace through forced cooling. The probe is designed to come into contact with the molten metal and the oxide layer above it repeatedly without suffering from the crippling freeze-up of the molten substances on the probe. Field trials at two different metal production shops in 1987 will conclude the first two phases of the feasibility study and a demonstration of the new methodology.

ACKNOWLEDGEMENTS

The author acknowledges many helpful inputs in defining the requirements for the sensor-probe by the members of the Collaborative Technology Unit 5-2 of the American Iron and Steel Institute. Able assistance in both experiment and numerical simulation by K. Lyu and J. Kralik is also acknowledged. This investigation is supported in part by the American Iron and Steel Institute CTU 5-2, the Ben Franklin Fund of the State of Pennsylvania, the Center for Metals Production, and Lehigh University.

REFERENCES

1. T.W. Harding and Y.W. Kim, "Direct Sampling of Gas and Particulates From Electric Arc Furnaces", in Physics in the Steel Industry, ed. F.C. Schwerer, American Institute of Physics, New York (1982), p. 362.

2. J.R. Porter, J.I. Goldstein, and Y.W. Kim, "Characterization of Directly Sampled Electric Arc Furnace Dust", in Physics of the Steel Industry, ed. F.C. Schwerer, American Institute of Physics, New York (1982), p. 377.

3. G.B. Rybicki and A.P. Lightman, Radiative Processes in Astrophysics, John Wiley and Sons, New York (1979).

4. Y.W. Kim, H.S. Lee and P. Sincerny, "Time-Resolved Temperature Measurement for a Pulsed Hydrogen Plasma Confined by Ultra Dense Hydrogen", in Proceedings of XVI International Conference on Phenomena in Ionized Gases, Dusseldorf (1983), p. II-256.

5. Y.W. Kim and P. Sincerny, "Reversal Line Profile of Copper Resonance Lines in Non-Ideal Hydrogen Plasmas", in Proceedings of XVII International Conference on Phenomena in Ionized Gases, Budapest (1985), p. 1001.

MACROSTRUCTURAL ENGINEERING OF PIEZOELECTRIC TRANSDUCERS

M. Kahn, B. Kovel and D. Lewis

Naval Research Laboratory
Washington, D.C. 20375-5000

Abstract

Dense piezoelectric lead zirconate titanate (PZT) exhibits a low hydrostatic pressure sensitivity. Application of the Gibbs thermodynamic energy function relates the piezoelectric response primarily to the Poisson's ratio. A moderate reduction of the Poisson's ratio will result in a significantly more compressible body, and its hydrostatic sensing ability will be greatly increased.

A sealed, PZT ceramic-air composite, used in the place of solid PZT will allow for greater compressibility and accordingly, greater hydrostatic response. The air phase of the composite is introduced by the rise of tape casting technology and photolithic methods, and the preparation procedure is discussed. The physical characteristics of void size and porosity are shown. Piezoelectric stress coefficients and dielectric constants as well as free field measurements of the sintered composites are also reported.

Introduction

The hydrostatic-pressure charge-sensitivity of conventional piezoelectric lead zirconate titanate (PZT) is only a fraction of its biaxial pressure sensitivity. Causes for this shortcoming are analyzed and a high hydrostatic gain ceramic-air composite structure is described.

Theory

Atomistic considerations suggest that the generation of piezoelectrically induced charges is a result of pressure-induced strain in the material. The piezoelectrically induced charge (Q) can then be calculated from the strain (ε), the stress coefficient (d), and the modulus (C).

$$Q_i = d_{imn} \, C_{mnkl} \varepsilon_{1} K_{1} \tag{1}$$

In a unit cell of PZT material as shown in Fig. 1, hydrostatic pressure induces the compressive strains ε_3, ε_1, and ε_2 as well as extentional strains ε_{31} and ε_{32}. The last two oppose strain ε_3 so that the strain under hydrostatic conditions becomes

$$\varepsilon_h = \varepsilon_3 + \varepsilon_{31} + \varepsilon_{32} \tag{2}$$

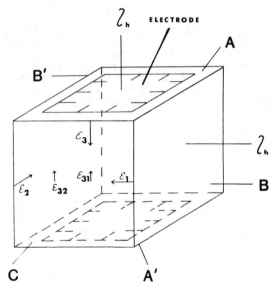

FIGURE 1: Hydrostatically Stressed PZT Ceramic

In an isotropic medium

$$\varepsilon_3 = \varepsilon_2 = \varepsilon_1 \tag{3}$$

Poisson's ratio is given by

$$\gamma = \varepsilon_{31}/\varepsilon_1 = \varepsilon_{32}/\varepsilon_2 \tag{4}$$

so that

$$\varepsilon_h = \varepsilon_3 + \gamma\varepsilon_1 + \gamma\varepsilon_2 \qquad (5)$$

giving

$$\varepsilon_h = \varepsilon_3(1 + 2\gamma) \qquad (6)$$

The hydrostatic piezoelectric response coefficient has been expressed[1] as

$$d_h = d_{33} [1 + (2d_{31}/d_{33})] \qquad (7)$$

If one then equates the hydrostatic correction factor of the strain in Eq. (6) to that of the piezoelectric response in Eq. (7), one gets

$$2\gamma = 2d_{31}/d_{33} \qquad (8)$$

This (somewhat heuristic) argument is also verified by the phenomonological Gibbs function for PZT[1][2] and proven further by the experimental results shown in Fig. 2.

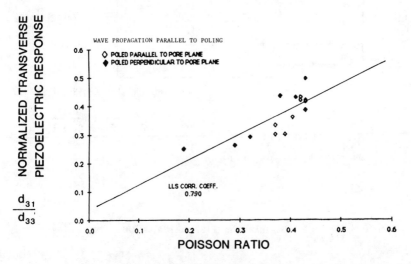

FIGURE 2: Normalized Transverse Piezoelectric Response Versus Poisson's Ratio Of PZT With Anisotropic Porosity

Background

PZT ceramic has a very high biaxial pressure response (d_{33}) but its Poisson's ratio is often above 0.4. Since the hydrostatic response

$$d_h = d_{33} (1-2\gamma) \qquad (9)$$

(from Eq. 7 and Eq. 8), one obtains $d_h < 0.2d_{33}$, a significant deterioration in sensitivity.

A lower Poisson's ratio means the body is more compressible. The high piezoelectric output of PZT material is critically dependent upon is chemical composition and its molecular structure. A high compressibility, high gain piezoelectric therefore requires a composite configuration that contains a specific phase of pure PZT. This compound cannot be adulterated by diffusion, e.g. from other materials added, during the high temperature used to form the PZT. It appears furthermore desirable that the second phase of the composite have maximum compressibility. To obtain high gain in a piezoelectric composite it is furthermore required that the PZT phase have a direct physical or electrical connectivity between the active sensing surface and the backing of the device. Any intervening insulating layer can be expected to have a potential across it that reduces the effective field across the PZT phase during poling and that lowers the available output voltage when a stress is applied to the composite. (See Fig. 3.)

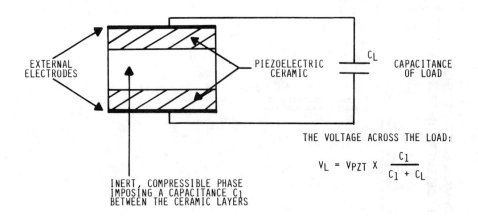

THE VOLTAGE ACROSS THE LOAD:

$$V_L = V_{PZT} \times \frac{C_1}{C_1 + C_L}$$

FIGURE 3 - SCHEMATIC OF PIEZOELECTRIC COMPOSITE: SERIES CONFIGURATION

A parallel PZT-polymer composite can be considered to have an enhanced compressibility (i.e., a lower Poisson's ratio), provided that the compressive force on the whole major surface of the composite is concentrated on the piezoelectric columns. This requires a stiff pressure plane over the composite.

114

The pressure, i.e. the stress, on the piezoelectric ceramic is then amplified, causing an increase in strain and in output voltage. This then appears as an increase in compressibility of the composite (see Fig. 4). On the other hand, the lateral (X-Y) pressure remains equal to the hydrostatic pressure. As a result the hydrostatic response of the composite is significantly above that of single phase PZT. The charge generated by a piezoelectric material decreases with a reduction in cross-sectional area, but it increases with an increase in pressure. Thse opposite trends make the piezoelectric output independent of the fraction of the area of the dense piezoelectric material in the sample (within limits of linearity) and permits the construction of relatively high hydrostatic gain, parallel composites.

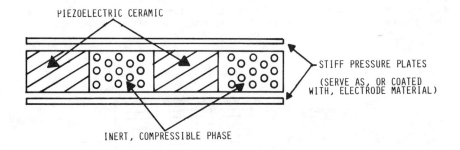

PRESSURE ON CERAMIC = APPLIED PRESSURE x $\dfrac{\text{AREA OF PRESSURE PLATES}}{\text{AREA OF CERAMIC}}$

FIGURE 4 - Schematic of Piezoelectric Composite: Parallel Configuration

Assembly of Discrete Parts. A relatively successful piezoelectric composite can be made by a parallel array of piezoelectric discs or rods in a polymer matrix, similar to what is shown in Fig. 4. The stress plate serves as a pressure amplifier and also collects the charge generated. The main disadvantage of such a configuration is the compressibility of the polymer is lowered as the static pressure is raised. This decreases the hydrostatic sensitivity of the composite when it is used in deeper water. The preparation and simultaneous assembly of many aligned piezoelectric rods and discs can also constitute a burden on the manufacture of such transducers.

Integral Ceramic Composite Structure. The advantage of using a fully sintered ceramic body is in its resistance to high temperatures and pressures, its imperviousness to water, its ease of electrode application and its relative rigidity. The latter minimizes spurious outputs due to excessive flexing. These properties are best utilized in a ceramic composite in which the compressible phase is fully encapsulated, so that the composite

presents a dense, sealed and rigid surface to the environment. Air, as the
second phase, has the advantages of a very high compressibility, not causing
contamination to the ceramic and being amenable to easy production into
ceramic. The rest of this presentation describes some of the preparation
techniques and properties of sealed ceramic-air composites.

Preparation of Ceramic-Air Composites

Figure 5 shows a model of a ceramic body, containing flat, uniformly
shaped and dimensioned voids, arranged in regularly spaced horizontal
planes.[3] The individual voids are aligned in ordered vertical columns.
Ceramic tape technology combined with the topical application of a fugitive
ink[1] as is used for impregnated electrode multilayer capacitors[5], was found
to be a suitable medium for making such a structure. To implement it, an ink
deposition screen is masked off with a pattern that is then used to deposit a
fugitive ink onto green PZT 5A ceramic tape. Tapes with these patterns are
then stacked and laminated under elevated pressures and temperatures.

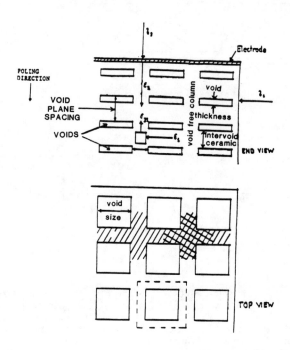

FIGURE 5:   Model Of Ceramic With Rectangular Voids

Upon burn-out, the organic constitutents of the tape, as well as those
of the ink, vaporize and diffuse out, leaving voids where the ink had been.
At the subsequent sintering, the ceramic around these voids, as well as that
at the edges, consolidates to a nearly fully dense state. The resultant
ceramic-air composite is then nearly impervious. When it is subjected to
hydraulic pressure, there is practically no increase of the gas pressure
within the pores.

116

Results and Discussion

Void Characteristics. Figure 6 shows a cross section of a ceramic sample with corner connected square shaped voids and Figure 7 shows a ground top surface of this sample. Figure 8 is the top view of a sample with unconnected square-shaped voids. Figure 9 shows the surface of a sample with disc-shaped voids.

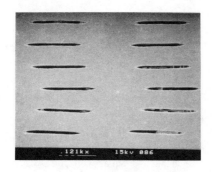

FIGURE 6.
Cross Section of Ground Ceramic
with Square Voids
Void Thickness 0.54 x 10-3 in.
Void Plane Spacing 3.7 x 10-3 in.
Void Length and Width 11 x 10-3 in.

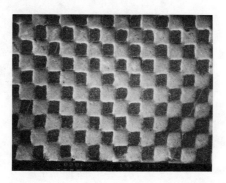

FIGURE 7.
Top View of Ground Ceramic Having
Corner Connected
Square Shaped Voids
Void Length and Width 11.5 x 10-3 in.

FIGURE 8.

Top View of Ground Ceramic Having
Unconnected Square Shaped Voids

Void Length and Width 29 x 10-3 in.
Void C-C Spacing 36.6 x 10-3 in.

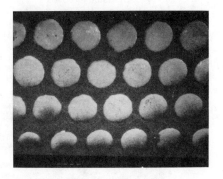

Top View of Ground Ceramic Having
Disc Shaped Voids

Void Diameter 24 x 10-3 in.
Void Surface C-C Spacing 30 x 10-3 in.

The samples without voids had 2%-5% of total porosity and about 0.5% of open porosity. The total porosity of the samples with internal void patterns was 11.6% to 21.6%. In general, the total porosity did not vary from sample to sample within a particular pattern.

The samples with voids exhibited some open porosity. This is attributed to defects introduced during sample preparation, e.g. the unintentional presence of voids at the surface.

Piezoelectric Stress Coefficients and Dielectric Constants. In order to determine the optimum poling voltage for ceramics with ordered voids, the samples were poled at successively higher voltages, up to 3.3 KV/mm, and measured after each step.

As Figure 10 shows, the d_{33} values initially increased rapidly with poling voltage. This increase became more gradual above 1.5 KV/mm. After poling at 2.1 KV/mm, the void-free blanks had a d_{33} of 483 pC/N. The samples with void patterns had d_{33} values that were 13% to 35% lower than those of the blanks, see Table 1.

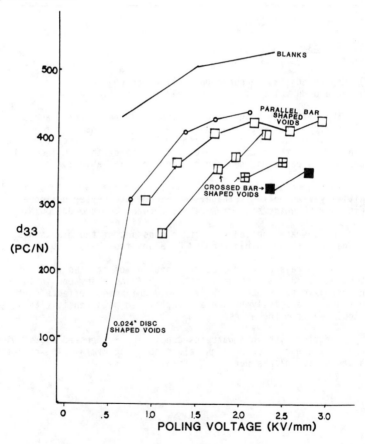

FIGURE 10: d_{33} Versus Poling Voltage

119

Table 1

Dielectric and Piezoelectric Pressure Response of PZT with Ordered Voids

Void Configuration	K Unpoled	Poled / K Unpoled	R (1)	d_{33}(2) (pC/N)	g_h (mVm/N (3)	d_h (pC/N)
Blank Laminates	1183	1.82	-0.4	483		
Crossed Bars 0.024" wide 0.010" spaced	210	1.22	-0.05	310	120	223
0.024" wide 0.024" spaced	386	1.19	-0.11	345	64	220
0.010" wide 0.010" spaced	454	1.09	-0.09	351	58	231
0.024" dia Disc (1.25" sq. plate)	447	1.25	-0.15	359	40	197

All poling at 2.1 KV/mm.

1) $R = \dfrac{d_{31} + d_{32}}{2d_{33}}$

2) d_{33} and d_{31} as measured on Berlincourt d_{33} meter.
3) g_h and d_h values from free field hydrostatic response.

The $d_{31} + d_{32}$ $2d_{33}$ ratio in the void-free blanks ranged to -0.43 after poling with >2.1K/mm. The ($d_{31} + d_{32}$ $2d_{33}$) ratios of samples with the various void patterns ranged from -0.2 for samples with 0.024" dia. disc-shaped void pattern to below -0.09 for some of the samples with crossed bar-shaped void patterns, see the column for R in Table 1.

The d_h values derived from the d_{33} and d_{31} parameters also increased initially with poling voltage, see Figure 11. After poling at 2.1 KV/mm the samples with voids showed d_h values between 243 and 306 pC/N.

The average unpoled dielectric constant of the blanks was 1183; the laminates with voids had unpoled K values from below 350 to above 550.

Poling raised the dielectric constant of samples with voids only by about 9% to 38%, as compared with the 82% increase seen in blanks. Increases in poling voltages above 1.5 KV/mm had only minor effects on the dielectric constant. Relatively high piezoelectric voltage sensitivities, g_h, can be calculated from these results.

Ceramics with void patterns consisting of crossed 0.024" wide x 0.010" spaced bars had calculated g_h values above 85 mVm/N. The calculated g_h values of the blanks were <9.2 mVm/N.

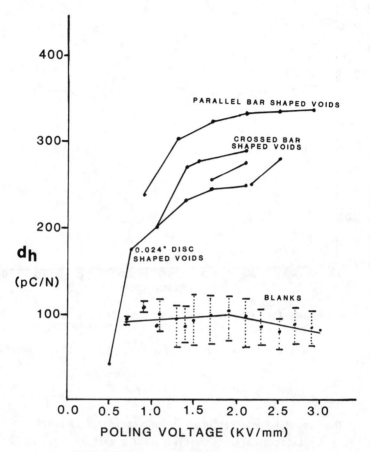

FIGURE 11: d_h Versus Poling Voltage

Free Field Hydrostatic Sensitivity. d_h and g_h parameters also were more directly determined from free field measurements and these are shown in Table 1. A $d_h \cdot g_h$ product as high as 26800 fm^2/N can be derived from the measurements. The free field responses are (in five out of six samples) 19.2% to 24% lower than those derived from the d parameter measurements on the Berlincourt d parameter tester. The cause for this is not yet completely clear, but otherwise the two measurement techniques seem to track quite closely.

The frequency response of these plates between 10 KHz and 100 KHz is shown in Figure 12 to be relatively flat.

For testing under higher pressures, samples with voids were encapsulated in a flexible epoxy. Preliminary tests indicate no change in hydrostatic sensitivity of these samples at pressures of 100 psi and no irreversible parameter changes under pressures in excess of 2000 psi. These investigations are proceeding.

121

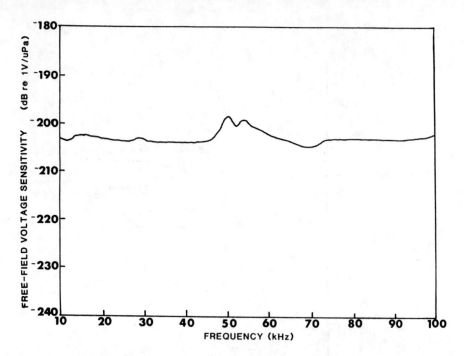

FIGURE 12: Free Field Open Circuit Voltage Sensitivity Of
1.25" Square Plate With 0.024" Diameter Disc Shaped Voids

(Sensitivity Level Shown Includes 4.2dB Of Cable Attenuation)

Summary

The high Poisson's ratio of piezoelectric PZT ceramic causes it to have
a low hydrostatic pressure sensitivity. The latter is raised through the use
of a ceramic-air composite structure in which a highly compressible second
phase (the air in the ordered voids) gives the composite a lower effective
Poisson's ratio. The integral ceramic composite structure containing ordered
columns of air-filled voids was made by using ceramic tape techniques and
photolithographically defined deposition of fugitive ink. A nearly flat
frequency response was observed between 10 kHz and 100 kHz, and the figure
of merit, $g_h \cdot d_h$ products were above 25,000 $\frac{fm^3}{N}$.

Acknowledgements

The support of this work by the Office of Naval Research and the
performance of the free field measurements by C. Le Blank at NUSC and by W.
Thompson at NRL are gratefully acknowledged.

122

References

1. M. Kahn, " Acoustic and Elastic Properties of PZT Ceramics with Anisotropic Pores", Jour. Am. Cer. Soc.68, No. 11, 623-628, 1985.

2. A.A.H. Amin, "Phenomological and Structural Studies of $PbZrO_3-PbTiO_3$ Piezoceramics", pp. 26, 33; Ph.D. Thesis, Pennsylvania State University, Nov. 1979.

3. "Preparation and Performance of Ceramic-Air Composites for Hydrostatic Sensing", M. Kahn, A. Dalzell and B. Kovel, to be published by the Am. Cer. Soc., 1986.

4. "Ceramic Matrices for Electronic Devices and Process", T. Rutt, J. Stynes, U.S. Patent #4353957, 1982.

5. "Multilayer Ceramic Capacitors with Injected Electrodes", N. Kenney, Workshop on the Reliability of Multilayer Ceramic Capacitors, NMAB-400, National Academy Press, pp. 285-292, 1983.

Process Modeling

COMPUTER MODELING OF SUPERALLOY CASTING

A. F. Giamei* and R. J. Mador**

*United Technologies Research Center, Silver Lane, East Hartford, CT 06108

**Pratt & Whitney, Manuf. Div., Main St., East Hartford, CT 06108

Abstract

Over the past fifteen years, several different methods have been used to calculate heat flow patterns during the solidification of superalloys. The calculations have included analog, analytical, finite difference and finite element methods. Applications have included 1D, 2D and 3D shapes ranging from equiaxed castings to directional structures. This latter category would include columnar grain airfoils, single crystal turbine blades and eutectic castings. Heat transfer by conduction, convection and radiation have been considered. This paper, however, will concentrate on FEM results from the directional solidification of constant and variable cross-section rectangular bars. These are three dimensional transient analyses with moving boundary conditions where the primary means of heat transfer is via radiation exchange within a vacuum chamber. Temperature dependent material properties were used and the heat of fusion was included.

Introduction

Advanced nickel-base superalloys are typically cast in vacuum chambers with long exposures at high temperatures. The ultimate in performance is achieved when the grain structure is controlled in the required fashion, e.g. a single crystal (Ref. 1). This implies the need for control of heat flow during the solidification process. Here, the geometries of interest are complex airfoils attached to root sections of very different cross sections. Such a configuration is shown schematically in figure 1. The solidification problem of interest is typically of the transient type. It is difficult to monitor the solidification sequence due to the large number of thermocouples which would be involved. In previous studies, many of the data reduction schemes for obtaining solidification parameters from temperature measurements (as well as computer simulations) have assumed one dimensional heat flow to simplifiy the problem (Ref. 2).

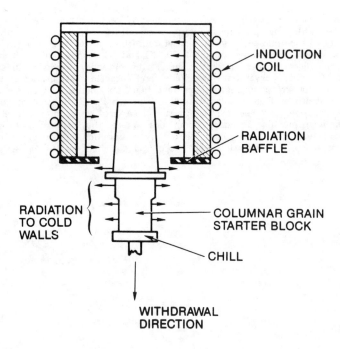

Figure 1 - Schematic of Directional Solidification Process.

128

The proper types of dendritic monocrystal microstructures can be generated in cylindrical shapes and this is where the bulk of the development and simulation work has been centered in the past. In order to build confidence in the simulation model, a concerted effort was started to transition from cylinders to stepped cylinders to rectangular bars to stepped rectangular bars to airfoils. This would allow for some comparisons with analytical results as well as corroboration with experimental data gathered along the way. The experimental data is acquired by monitoring thermocouples inserted along the centerline of a centrosymmetric casting. In addition, thermocouples may be present just outside the mold or at the shell mold/metal interface, preferably at the same horizontal position as the interior thermocouples. This allows for the computation of axial and radial thermal gradients. The axial thermal gradients determine the solidification morphology (Ref. 3,4) whereas the radial thermal gradients control the evolution of extraneous grain nucleation and feeding/shrinkage.

The EMF from thermocouples along a vertical line are monitored as a function of time. These data are fed into a data reduction program which plots cooling curves, curve fits temperature vs. distance, computes interface position as a function of time for liquidus and solidus, derives thermal gradient values vs. time for both interfaces, curve fits interface position vs. time, computes growth rates for the two isotherms of interest, cross-plots growth rate vs. thermal gradient, computes mushy zone height vs. time and calculates local solidification times. All of these quantites are plotted in convenient form. A completely analagous set of (simulated) solidification parameters can be obtained from the predicted thermal behavior during the thermal transient.

Computational Approach

The general computational approach is described in some detail elsewhere (Ref. 5). FEM was selected (Ref. 6) due to the greater degree of freedom in mesh design and computational accuracy, as compared to FDM. Some debit was anticipated in terms of run time. The computations were carried out on a VAX 11/785 (rectangular bar) or a VAX 8600 (stepped rectangular bars) using the commercially available K2 version of the MARC FE, non-linear heat transfer computer program. This program allows for such non-linearities as temperature dependent material properties (including phase change) and time-dependent radiation boundary conditions. Pre- and post-processing were carried out using the commercially available PATRAN program. The boundary conditions were moved upward along the bars to simulate the withdrawal of the mold from the hot zone to the cold zone in the real world.

The shell mold thickness was taken to be a nominal 0.25 in (0.64 cm). For the constant cross-section case the section size was 0.375x1.0 in (0.95x2.54 cm) giving a quadrant for meshing of 0.188x0.5 in (0.48x1.27 cm); the length was 8 in (20.3 cm). For the cases with a transition in section size, the lower and upper ends were each 3.75 in (9.52 cm) long and the section size for the smaller area was 0.25x0.75 in (0.64x1.91 cm) giving a quadrant size of 0.125x0.375 in (0.32x0.95 cm); for the larger area, the larger section size was 0.75x2 in (1.91x5.08 cm) and the quadrant size was 0.375x1 in (0.95x2.54 cm). The mesh dimensions were either 0.125 or 0.25 in (0.32 or 0.64 cm) in x, y, or z depending on whether the element was in the shell or metal and whether the mesh location was close to a transition. A conforming baffle was placed in the model between the hot zone and cold zone regions. The baffle clearance was 0.125 in (0.32 cm) in all the radial directions (before and after changes in cross section). View factors of the hot zone, baffle region and cold zone were computed analytically for each geometry which was considered.

A linear shape function was selected and the validity of the mesh selection was determined by examination of the convergence of solutions for other element sizes as well as higher order approximations. Reasonably fine mesh sizes were used, particularly in the metal near any transitions in cross-section. Adaptive time steps were controlled in order to avoid instabilities and yet to minimize computer run time. The computations reviewed here took 3-6 hours of VAX 8600 CPU time. (This is a considerable improvement over the several weeks to months required to make wax injection tools, create wax injections, assemble with appropriate gating to make a casting, dip, dry, dewax, cast, solidify, remove from mold and inspect for quality in the real world. In addition, much more information about the thermal events which "occurred" are available at very low cost.)

Temperature dependent material properties were taken from published values (Ref. 7). The heat of fusion was incorporated into the specific heat function as a triangular spike which reached a maximum value midway through the freezing range. Although high strength superalloys tend to freeze more just below the solidus than elsewhere in the freezing range, previous studies has shown the symmetric case was a good approximation for the purposes intended here. (Adaptive time steps were used to ensure that there was no computational loss of latent heat enthalpy). An initial uniform temperature of 2750°F (1510°C) was assumed in the melt at the beginning of withdrawal, which was assumed to be at a constant 6 in/hr (15.2 cm/hr). A withdrawal transient was computed out to 5000 sec (8.3 in or 21.1 cm withdrawal distance).

Results

There are many ways in which the results can be presented. The data are available as arrays of temperature vs. both time and distance. Obviously, the raw data are very detailed and difficult to assimilate in a short period of time. The data can be reduced (much like actual thermocouple data) to form line drawings, e.g. of the dependence of the relevant solidification parameters on time or distance solidified. These curves are most useful when only approximate numerical values are required or when computed quantities are to be compared to measured values. The next level of complexity would be a 2D view (i.e. on a planar section) of the 3D body. Isotherms can be shown as line contours or as colored patches. Finally, and isometric view can be taken at any point in time for various "magnifications". The isometric view combined with color graphics is most useful in obtaining a quick feel for the problem. These views can then be assembled into a video to envision the time dependence. For this paper, we will concentrate on the reduced data curves and the isometric views of the casting at a given point in time.

Figure 2 shows a closeup isometric view of the constant area rectangular bar at 2180 sec. The baffle is included and can be used as a scale factor for the bar and the elements since it was 0.31 in (0.79 cm) thick. The view is from the centerline outward. (Actual post-processed results from PATRAN are typically exhibited on interactive screens or on color plotters as images with several colors and shades available for each "snapshot". The color coding can be seen to be repetitive in black and white and was only selected for maximum contrast. The "color" coding for all figures is shown on a color bar to the right and runs from 1000 to 2750°F (538 to 1510°C). The solidification range (2360 to 2540°F or 1294 to 1394°C) is seen to be just above the baffle, i.e. in the hot zone. This is consistent with past results for tightly baffled simple shapes withdrawn at low velocity, which have been characterized as associated with a heat input limited process. Note that the isotherm curvature within the solidification region is concave down, i.e. the outside is hotter than the inside. This is consistent with the notion that in this process, the heat flows radially inward in the hot zone, downwardly through the body near the baffle and radially outward in the cold zone with the isotherms generally perpendicular to the lines of heat flow. The axial thermal gradient in the mushy zone is approximately 500°F per in (111°C per cm), and is consistent with past measurements made under these conditions. Ideally, the interfaces are centered in the baffle, the solidification isotherms are flat and in a region of high axial thermal gradient and low radial gradients within the metal. The case analyzed here is close to ideal. It would suggest that the withdrawal rate could be increased for even better conditions. This is again consistent with practice which suggests that the optimum withdrawal rates for simple shapes is 2.5X faster than the rate selected for analysis here (more typical of rates required for complex shapes).

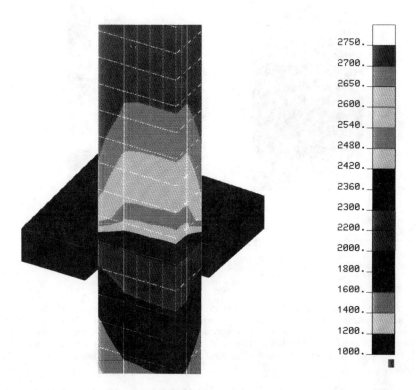

Figure 2 - Closeup isometric view from centerline of rectangular bar at 2180 sec. Baffle position is illustrated.

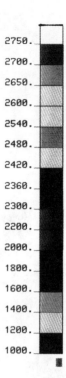

Figure 3 - Same case as figure 2 but with view of metal/mold interface from the inside.

Figure 3 shows an isometric closeup view for the same body and transient time as in figure 2, but in this case the view is of the mold/metal interface from the metal outwards. The baffle is not shown here, but is positioned at the same location as in figure 2. Note that the centerplane mold temperatures are identical to those shown in figure 2 since they represent the same surfaces at the same time.

Figure 4 shows a comparison of experimental data (points) vs the corresponding computed cooling curves for the rectangular bar. These data exhibit the temperature at the midpoint of the wide face side of the mold/metal interface. The agreement is reasonable, particularly at the higher temperatures. The measurements at the lower temperatures seem to lag. As previously discussed in Ref. 5, these discrepancies are primarily caused by thermocouple conduction errors. In addition, it may be important to recognize that at least in simple shapes the thermal contraction on cooling is likely to cause separation or "break-away" leading to a gap between mold and metal in the coolest region, which decreases the rate of cooling. Such a "break-away" event was not taken into account in the current work.

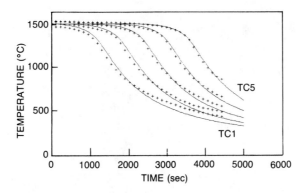

Figure 4 - Computed (lines) vs. measured (points) cooling curves for the rectangular bar along the midpoint of the wide side of the metal/shell interface.

For the transitions to be discussed below, however, it was felt that the most meaningful comparisons of solidification would be obtained along the centerline. The computed data for the rectangular bar was processed as outlined above to obtain the freezing curves shown in figure 5. As one might expect, a relatively steady process is indicated by the nearly linear dependence of liquidus and solidus interface positions on time. The slopes of these curves correspond well with the withdrawal rate. The baffle center location is zero at time zero and moves along the bar at 6 in/hr (15 cm/hr). Superposition of a straight line representing the baffle center location onto figure 5 confirms that the solidification interfaces are above the baffle throughout the withdrawal transient.

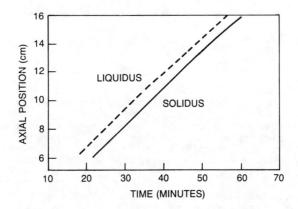

Figure 5 - Solidus and liquidus axial position vs. time at the centerline of the constant cross section rectangular bar.

133

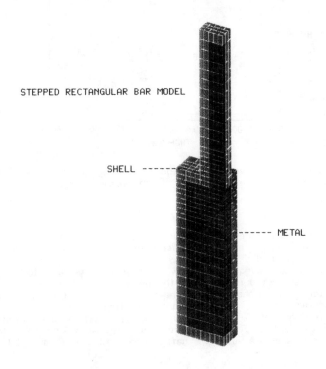

STEPPED RECTANGULAR BAR MODEL

SHELL ------

------ METAL

Figure 6 - FEM model for stepped rectangular bar.

Figure 6 shows the model used for the stepped rectangular bar model with large end down (i.e., withdrawn first). Again the view of this quadrant model is from the center outward. Note that the mesh is finer in the metal than in the mold and much finer near the transition. Figure 7 shows a closeup isometric view of this geometry 1894 sec into the withdrawal cycle. The view is from the center outward. Note that again, the axial gradients are high and the radial gradients are low within the metal. The solidification region is above the baffle (not shown). At longer times, e.g. 2133 sec as in figure 8, the solidification isotherms begin to converge and become slightly irregular in shape. As solidification proceeds the solidus and liquidus isotherms climb further up into the hot zone. This reults from the existence of a higher surface to volume ratio in the upper (smaller) section of the casting, which leads to increased radiation leakage (as a percentage of total heat content) through the baffle opening. Thus an acceleration in growth velocity above the withdrawal rate would be expected until steady state is restored at some finite distance beyond the section change.

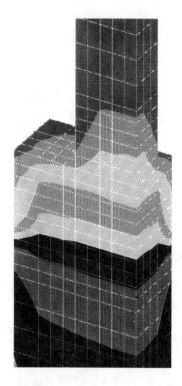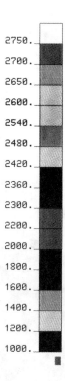

Figure 7 - View of large end down stepped rectangular bar from center outward at 1894 sec.

Figure 9 shows the companion view of the metal/mold interface. The x-y-z axis label is a reminder that in this case the view is from underneath whereas the previous views were from above. The z axis goes into the paper. The corresponding freezing curves are shown in figure 10. The solidus surface is shown to be advancing at a rate faster than the withdrawal rate from 25 min to 50 min which is when the transition in geometry is near the baffle. The solidus surface first advances more rapidly than the withdrawal rate and then decelerates, leading to an improvement in axial thermal gradient as time proceeds. (The axial thermal gradient is proportional to the vertical separation of the curves on this plot.)

It is interesting to note that the calculated effect of this transition is exactly opposite that of a similar transition in geometry but with a convective bath used for heat removal (Ref. 2). In the earlier work a "wiper" baffle was used and the radiation leakage was not even a consideration. In the present case, the transition leads to an acceleration of solidification relative to the withdrawal rate, whereas with tight baffling there was a deceleration based on computation and actual microstructural observations.

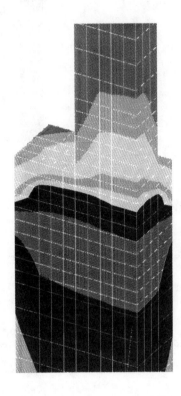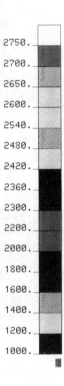

Figure 8 - View of large end down stepped rectangular bar from center outward at 2133 sec.

The last case to be considered is the transition from a large section to a small section. The same model was used as for the previous case but the moving boundary conditions were forced to move in the opposite direction and the "hot" and "cold" zones were reversed. The temperature distribution at 2248 sec into the transient is shown in figure 11. The view is from the centerline outward and again the perspective is from underneath. The corresponding computed freezing curves are shown in figure 12 and depict the least steady of the three geometries considered here. The more significant radiation leakage which dominates the first 2000 sec of the transient causes both liquidus and solidus to decelerate with time in the vicinity of the section change. As the larger section approaches the baffle, the influence of leakage rapidly diminishes and the solidification is retarded.

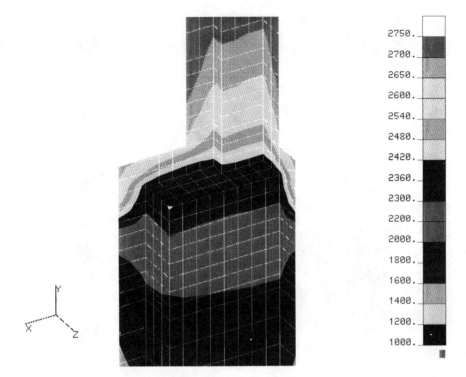

Figure 9 - As in figure 8 but view is of the metal/mold interface from the metal side looking upwards into the step.

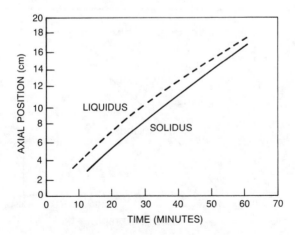

Figure 10 - Freezing curves along the centerline of the stepped rectangular bar with large end down.

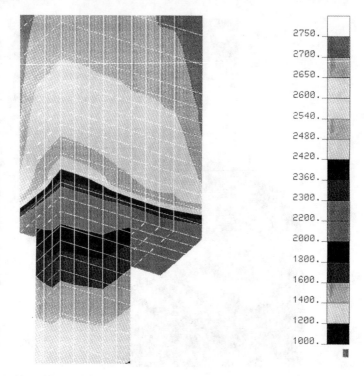

2750.	
2700.	
2650.	
2600.	
2540.	
2480.	
2420.	
2360.	
2300.	
2200.	
2000.	
1800.	
1600.	
1400.	
1200.	
1000.	

Figure 11 - View of stepped rectangular bar (with large end up) from center outward at 2248 sec.

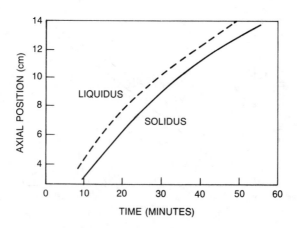

Figure 12 - Freezing curves along the centerline of the stepped rectangular bar with large end up.

Summary

Finite element models were constructed for a rectangular bar, a bar with a transition from large to small section size and vice-versa. Directional solidification of a superalloy within a shell mold was simulated by FEM analysis. The data were reduced both in graphical form and to yield freezing curves. Experimental data were gathered in the case of the rectangular bar for an identical shape, and the same materials and temperatures used in the simulation. The constant cross section model showed a steady state process. The transition from large to small caused a slight acceleration in solidification rate relative to the withdrawal rate. The transition from small to large section size led to the greatest deviation from steady state with a significant decrease in growth velocity in the vicinity of the geometrical transition. These results can be reconciled in terms of the influence of radiation leakage from the mold in the hot zone, through the baffle and into the cold zone.

Acknowledgement

The authors would like to thank M. Duffy and T. Holland of Pratt & Whitney, Manufacturing Division, for their computational assistance and D. Price and W. Gustafson of Pratt & Whitney, Engineering Division, for their assistance in obtaining the experimental data. We are also thankful for the many helpful discussions with Prof. F. Landis of U. of Wisconsin.

References

1) "The Development of Single Crystal Superalloy Turbine Blades," Superalloys 1980, ed. J. Tien et al. (Metals Park, OH: American Society for Metals, 1980), 205-214.

2) "Computer Applications in Directional Solidification Processing," Superalloys, Metallurgy and Manufacture, ed. B. Kear et al. (Baton Rouge, LA: Claitor's, 1976), 405-424.

3) "Manipulation of Superalloy Microstructures and Properties by Advanced Processing Techniques" Fundamental Aspects of Structural Alloy Design, ed. R. Jaffee and B. Wilcox (New York, NY: Plenum, 1977).

4) "The Art and Science of Unidirectional Solidification," New Trends in Materials Processing, ed. C. Hartley, T. Tietz and B. Kear (Metals Park, OH: American Society for Metals, 1976), 48-97.

5) "Finite Element Simulation of the Directional Solidification Process for Cylinders (2D) and Rectangular Bars (3D)," Modeling and Control of Casting and Welding Processes, ed. S. Kou and R. Mehrabian (Warrendale, PA: The Metallurgical Society, 1986), 433-448.

6) O. C. Zienkiewicz, The Finite Element Method (London, England: McGraw-Hill, 1977).

7) T. K. Pratt, F. Landis and A. F. Giamei, "Iterative Nonlinear Thermal Analysis of Directional Solidification Using a Fast Elliptic Solver," Numerical Heat Transfer, 4(1981) 199-213.

139

ESTIMATION MODEL OF AUSTENITE RECRYSTALLIZATION DUE TO

PLATE ROLLING AND ITS APPLICATION TO CONTROLLED ROLLING

A. Yoshie, H. Morikawa, Y. Onoe and K. Itoh

R & D Lab. II, Nippon Steel Corporation,
1-1-1 Yawata Higashi, Kitakyushu, 805 JAPAN

Abstract

The behaviors of static recovery, recrystallization and the retained strain of austenite (γ) in plate rolling process have been formulated in terms of the change of average dislocation density calculated from the decrease in the stress observed by the double deformation tests. By the present formulation, deformation stresses of various deformation conditions and the critical conditions of static recrystallization can be calculated as a function of temperature, strain, strain rate, γ grain size and interval time between the successive rolling passes of plate. Calculated results reveal that deformation conditions affect the critical condition of static recrystallization. The smaller interval time extends the no recrystallization temperature range to the higher temperature side.

The relation between the calculated retained strain due to controlled rolling and the mechanical properties of the plates has been also investigated. The increase in retained strain just before γ/a transformation improves the toughness of the plate. In order to increase the retained strain by the control of the progress of recovery and recrystallization, it is important to keep the rolling temperature in low temperature range and to finish the rolling in shorter period.

Introduction

Controlled rolling process has become one of the main processes in steel plate production in these days. Especially the total reduction in no recrystallization temperature range has been increased for the purpose of improving mechanical properties. Therefore, it is important for the process design to make clear the behavior of recrystallization and the critical condition for recrystallization of austenite (γ) in hot rolling process.

As seen in the schematic illustration, Fig. 1, the critical condition for static recrystallization of γ is considered to be a function of the temperature and the strain in the rolling as well as chemical compositions of steels. In the higher temperature range, the critical strain (ε_{cr}) is smaller than the strain of each rolling pass. As a result, recrystallization starts during the interval time before the next pass. In the lower temperature range, however, recrystallization does not start until the accumulated strain of several passes exceeds ε_{cr} because the strain of each rolling pass does not exceed ε_{cr}.

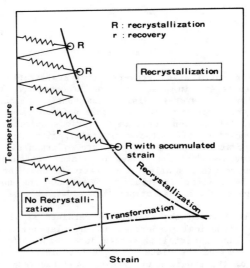

Figure 1-Schematic illustration of recrystallization of
γ in hot rolling process

The behavior of recrystallization has been investigated by many authors. Some authors[1]~[3] have observed the microstructure of the steel quenched after the rolling and shown the maps of the rolling conditions which are divided into recrystallization range and no recrystallization range. Other authors[4],[5] have performed the double deformation tests and expressed the progress of recrystallization as the change of the softening ratio which had been calculated from the decrease in the stress. The results of these reports might be applied to the rolling of 1 pass but they could not be applied to the practical multi pass rolling of plate because the strain accumulated in γ by the rolling in the no recrystallization temperature range is not taken into consideration. Recently some authors[6]~[8] have expressed the behavior of recrystallization in multi pass hot rolling process as a form of mathematical model by the formulation of the change of observed microstructure. But the critical condition for static recrystallization was not described mathematically, because incubation period of recrystallization was ignored in their calculation.

In the present study the behavior of static recovery and recrystallization in hot rolling process have been expressed in the form of mathematical
model by taking the incubation period into consideration. The model has been
derived from the decrease in the stress observed by double deformation tests.
Observed stress has been related to the average dislocation density because
the relations among dislocation, strain and stress have been already studied
by many authors.

Mathematical model

Procedure of mathematical analysis

Fig. 2 shows the schematic illustration of the change of the average
dislocation density and the stress in the double deformation test[9]. Average
dislocation density of γ increases when γ is deformed and decreases during
interval time due to the progress of recovery and recrystallization. On
assumption that stress (σ) corresponds to average dislocation density (ρ),
the value of ρ at each point (for example, point A or B in Fig. 2) can be
calculated from the observed σ (for example, stress strain curve A or B in
Fig. 2).

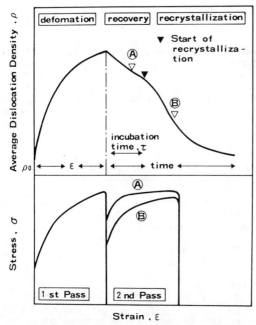
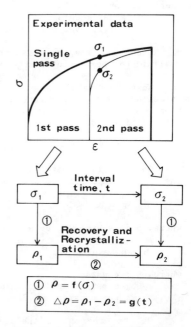

Figure 2–Schematic illustration of the
change of dislocation density
and stress strain curves

Figure 3–Schematic illustration of the proce-
dure of the analysis

Senuma et. al[10] indicated the method of calculating the deformation resistance of hot strip as a function of ρ. At the 1st step of the present
analysis, the relation between σ and ρ has been formulated as the same way.
At the 2nd step, the decrease in dislocation density due to recovery and
recrystallization has been formulated as a function of time after deformation
including unknown coefficients. At the 3rd step, the values of these unknown
coefficients have been decided by the data of double deformation tests from
which the decrease in dislocation density during interval time has been calculated. With these procedures, the progress of recovery and recrystalli-

zation after deformation have been formulated mathematically. Fig. 3 shows the schematic illustration of the procedure of analysis.

Relation among deformation stress, strain and average dislocation density

Deformation stress during hot deformation is expressed as[11] :

$$\sigma = \sigma_{id} + \sigma_e \tag{1}$$

where, σ_{id}; internal stress caused by interaction of dislocations.
σ_e; effective stress when dislocation gets over obstacles by thermal activation process.

The second term, σ_e can be neglected because σ_e is much smaller than σ_{id} when γ is deformed at high temperature.[12] Consequently the relation between σ and ρ is expressed as[13]

$$\sigma = \sigma_{id} = a\rho^{1/2} \tag{2}$$

where, $a = \alpha\mu\beta$
where, α; constant
μ; shear modulus
β; Burgers-vector.

Since the temperature dependence of μ is considered to be relatively small in the temperature range of the present study, μ is set as a constant.

Next, the relation between strain (ε) and ρ has to be derived. Strain hardening and dynamic recovery proceeds at the same time during hot deformation. The change of dislocation density is formulated into the following differential equation.

$$d\rho = (\delta\rho/\delta\varepsilon)d\varepsilon + (\delta\rho/\delta t)dt \tag{3}$$

On assumption that ρ is proportional to ε[14] , the rate of strain hardening ($\delta\rho/\delta\varepsilon$) is expressed as

$$\delta\rho/\delta\varepsilon = b \tag{4}$$

where, b is assumed to be a function of temperature[12] as

$$b = b_0 \exp(Q_b/RT) \tag{5}$$

where, b_0; constant
Qb; apparent activation energy
R; gas constant ($8.314 \ J\cdot mol^{-1}K^{-1}$).

On the other hand, the rate of dynamic recovery is assumed to be expressed as[15]

$$\delta\rho/\delta t = -c\rho^n \tag{6}$$

where, c is assumed to be a function of temperature, strain rate ($\dot{\varepsilon}$) , γ grain size (D_γ) as eq. (7)

$$c = c_0 \ D_\gamma^{m_c} \ \dot{\varepsilon}^{n_c} \ \exp(-Q_c/RT) \tag{7}$$

where, c_0 , m_c and n_c; constants
Qc; apparent activation energy.

The effect of temperature is assumed to be expressed as the same type of

144

exponential function as that of the thermal activation process. The effect of $D\gamma$ and $\dot{\varepsilon}$ are assumed to be expressed in the form of power function because the dependence of them on c is not clear. On the same assumption, the similar function forms are adopted in eq.(10), (14) and (15). As seen in Table I, value of n in eq.(6) depends on the mechanism of recovery. The chance for annihilation of dislocations of opposite Burgers-vector is very small because dislocations in f c c crystal hardly cross-slip.[16] Therefore, dislocation annihilates due mainly to climbing and absorption at grain boundaries.

Table I. Mechanism of annihilation of dislocation

n	Mechanism
1	Climbing or Absorption at Grain Boundary
2	Annihilation of Dislocations of Opposite Burgers Vectors

With the combination of eq.(4) and eq.(6), assuming n = 1, the relation between ρ and ε is formulated under the condition of $\dot{\varepsilon}$=constant through the deformation as

$$\rho = b/c(1-e^{-c\varepsilon}) + \rho_0 e^{-c\varepsilon} \tag{8}$$

where, ρ_0 ; dislocation density of annealed steel.

By eq.(2) and eq.(8), stress strain relations of any deformation conditions of the controlled rolling of plate in γ temperature range can be calculated. Dynamic recrystallization and γ/a transformation are not considered in the present study because this mathematical model is aimed at the application to the controlled rolling of plate in γ temperature range.

Relation between static recovery and average dislocation density

The decreasing rate of ρ during static recovery after deformation is expressed as the same form function as that of dynamic recovery as[15]

$$\delta\rho/\delta t = -d\cdot(\rho-\rho_0) \tag{9}$$

where, t; time after deformation

$$d = d_0 \ D\gamma^{md} \ \dot{\varepsilon}^{nd} \ exp \ (-Q_d/RT) \tag{10}$$

where, d_0, m_d and n_d; constants
        Q ; apparent activation energy.

Under the conditions that $\rho=\rho_d$ at t=0 and $\rho=\rho_0$ at t=∞, the relation between ρ and t is formulated into

$$\rho = (\rho_d-\rho_0) \ exp(-d\cdot t) + \rho_0 \tag{11}$$

where, ρ_d; dislocation density obtained just after the deformation.

Relation between static recrystallization and average dislocation density

After static recrystallization begins at $t = \tau$ (τ; incubation period), ρ decreases according to eq.(12) due to the progress of recrystallization and becomes ρ_0 after recrystallization completes.

$$\rho = (\rho_0-\rho_r)X + \rho_r \tag{12}$$

145

where, X; fraction of recrystallized Y
  ρ_r; dislocation density according to eq.(11).

X is formulated as a function of t into following equation[17]

$$X = 1-\exp\{-e(t-\tau)^n\} \tag{13}$$

where, e is assumed as a function of T, ε, $\dot{\varepsilon}$ and $D\gamma$ as

$$e = e_0 \; D\gamma^{me} \; \dot{\varepsilon}^{ne} \; \varepsilon^{le} \; \exp(-Q_e/RT) \tag{14}$$

where, e_0, m_e, n_e and l_e; constants
   Q_e; apparent activation energy.

Incubation period, τ is defined as a time between the deformation and the first nucleation in eq.(13) and is also assumed to be a function of T, ε, $\dot{\varepsilon}$, and $D\gamma$ as

$$\tau = \tau_0 \; D\gamma^{m\tau} \; \dot{\varepsilon}^{n\tau} \; \varepsilon^{l\tau} \; \exp(Q_\tau/RT) \tag{15}$$

where, τ_0, m_τ, n_τ and l_τ; constants
   Q_τ; apparent activation energy.

Through this analysis n is assumed to be constant, 2 on the basis of the present experimental data and the report of other authors[7,18]

   With the combination of the equations mentioned above, static recovery and recrystallizaiton behaviors and the critical condition of recrystallization can be calculated. This mathematical model can be applied to multi pass rolling of plate because average dislocation density, ρ can be calculated consistently from the beginning to the end of rolling process.

Materials and experimental methods

   The chemical compositions of the steels used are shown in Table II. Deformation tests with single or multi passes were performed using a compression type hot deformation simulator[19]. Specimens were taken from continuous casting slabs and machined to the column which size is 7 mm diameter and 12 mm height. They were compressed to the axial direction. Strain, deformation speed, interval time during deformation and temperature were precisely controlled by computer. Specimens were heated to the heating temperature (HT) at the heating rate of 5°C/s, and held for 10 minutes followed by cooling to the deformation temperature (DT) at the cooling rate of 5°C/s. Then single or multi passes of deformation were performed at the temperature. In the case of double deformation test, the deformation stress of the 2nd pass decreased due to the progress of recovery and recrystallization according to the increase in the interval time.

Table II. Chemical compositions of the steels

Steel	C	Si	Mn	P	S	Cu	Ni	Cr	Aℓ	Nb	B	N
A	0.12	0.26	1.00	0.014	0.004	0.17	0.91	0.57	0.071	–	0.001	0.0045
B	0.07	0.23	1.33	0.016	0.004	0.46	0.79	–	0.037	0.01	–	0.0056

Results

Comparison between microstructure and deformation stress

The high hardenability steel A containing Ni, Cr and B was employed for the observation of the prior γ grain boundaries of a quenched specimen after deformation. Fig. 4 shows the experimental conditions. The prior γ grain boundaries were observed in the specimens quenched just after the holding time ranging 1s - 1000s after the 1st pass. Stress strain curves of the 2nd pass were measured using other specimens which were deformed just after the same holding time. Fig. 5 and Fig. 6 are the optical microstructures of the quenched specimen and the observed stress strain curves respectively. These figures show the decrease in the stress of the 2nd pass due to the progress of recrystallization. Fig.7 shows the relation among the times after the 1st pass, the fraction of recrystallized γ and the stress of the 2nd pass. X was measured by observation of microstructure and σ is adopted the stress at $\varepsilon = 0.05$ of the 2nd pass. Fig. 7 reveals that the decrease in σ for t < 10s was caused by recovery and that for t $\geq$ 10s was caused by recrystallization.

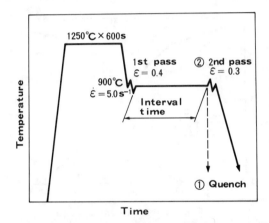

Figure 4-Experimental conditions for steel A

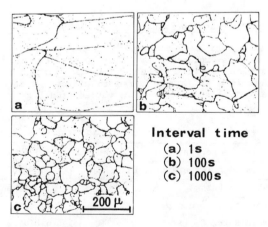

Interval time
(a) 1s
(b) 100s
(c) 1000s

Figure 5-Progress of recrystallization during interval time after the 1st deformation

147

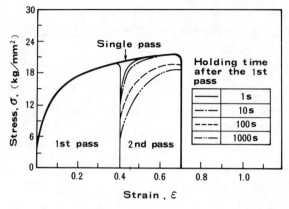

Figure 6-Effect of interval time on the stress strain curves

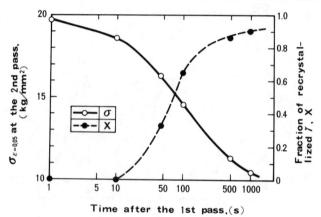

Figure 7-Effect of interval time on $\sigma_{\varepsilon=0.05}$ and X

Formulation of recovery and recrystallization behaviors of 0.01% Nb steel

The Nb steel B, which was the typical steel for the plate produced by controlled rolling, was employed. Table III shows the experimental conditions. Preliminary experiments confirmed that Nb had been completely dissolved even at HT=950°C. Therefore HT is considered to affect the stress only through the γ grain size. In this experiment, D_γ varied from 12.5μ to 300μ according to HT.

Table III. Experimental conditions for steel B

Heating Temperature, HT, (°C)	950,1050,1200
Deformation Temperature, DT, (°C)	750∿1000
Number of Passes	1, 2, 6
Deformation Strain per Pass, ε	0.2, 0.4, 1.2
Strain Rate, $\dot{\varepsilon}$, (s^{-1})	1.5,5.0,50.0
Holding Period during Passes, t, (s)	0.5∿1000

The possible effect of Nb precipitation on the stress by holding at certain temperature before deformation was also investigated. But its effect was very small because both C and Nb contents were relatively small in the steel B. Therefore precipitation hardening can be neglected and only strain hardening caused by retardation of recovery and recrystallization are considered to be an effect of Nb on the stress.

The constants included in equations (2), (5) and (7) were decided by regression analysis with the observed data of the stress strain curves. The units of observed data are kg/mm^2 for σ and μm for $D\gamma$. Table IV shows the values of constants. The effect of $D\gamma$, m_c and $\dot{\varepsilon}$, n_c on the rate of dynamic recovery are relatively small. The dependence of temperature on deformation resistance, which is usually expressed as Misaka's type equation[20], is divided into the dependence of strain hardening, Q_b, and that of dynamic recovery, Q_c. Their values are of the same order of magnitude but their effects are opposite. As a result, the stress increases with the decrease in deformation temperature.

By substitution of Eq. (13) into Eq. (12), the relation among ρ, e, t and τ is obtained. By expanding the term of t and τ into series and neglecting the terms of higher order after twice logarithmic calculation, it is possible to obtain a linear equation of ρ, e, t and τ. The values of e and τ are determined by regression analysis with the series of ρ calculated from the observed stress at $\varepsilon = 0.05$ ($\sigma_{\varepsilon=0.05}$) in the 2nd pass of double deformation tests in which T, ε, $\dot{\varepsilon}$ and $D\gamma$ are constant and t a variable. The constants included in Eqs. (14) and (15) are also obtained by regression analysis with the calculated e and τ for different T, ε, $\dot{\varepsilon}$ and $D\gamma$. The values of constants are also shown in Table IV. The effect of $D\gamma$, m_d and n_d are also small on the progress of static recovery. The effect of $\dot{\varepsilon}$, l_e on the progress of recrystallization is by far larger than that of $D\gamma$ and $\dot{\varepsilon}$. The value of l_e is close to that of experimental formula by other author[21].

Table IV. Constants for eq.(2),(5),(7),(10),(14) and (15)

eq	Coefficients	Constants	eq	Coefficients	Constants	eq	Coefficients	Constants
(2)	a	1.68×10^{-4}	(10)	d_o	8.78×10^7	(14)	m_e	-0.217
(5)	b_o	1.00×10^9	(10)	Q_d	201000	(14)	n_e	-0.254
(5)	Q_b	41300	(10)	m_d	-0.134	(15)	τ_o	1.40×10^{-12}
(7)	C_o	37.2	(10)	n_d	0.0772	(15)	Q_τ	241000
(7)	Q_c	17300	(14)	e_o	2.80×10^{20}	(15)	l_τ	-2.09
(7)	m_c	0.0412	(14)	Q_e	432000	(15)	m_τ	0.227
(7)	n_c	-0.0986	(14)	l_e	4.29	(15)	n_τ	0.159

Comparison between experimental data and calculated results

Fig. 8 shows the stress strain relations of various deformation conditions. Calculated results with equations (2)$\sim$(8) are in good agreement with experimental data in every case. Fig. 9 illustrates the decrease in the stress at the 2nd pass due to the progress of recovery after deformation. Calculation was performed on assumption that only recovery progressed. Experimental data are on the lines of calculated results before the start of recrystallization and part from them due to the progress of recrystallizaiton. In the case of the lower deformation temperature, Fig. 9(a), the incubation period for recrystallizaiton is expected to be between 20s and 100s for $\varepsilon = 0.2$ and between 10s and 20s for $\varepsilon = 0.4$. In the case of the higher deformation temperature, Fig. 9(b), recrystallization starts in much shorter period after deformation. The incubation period is expected to be between 1s and 2s for $\varepsilon = 0.2$ and within 0.5s for $\varepsilon = 0.4$. Fig. 9 reveals that the larger deformation strain and the higher deformation temperature result in the smaller incuba-

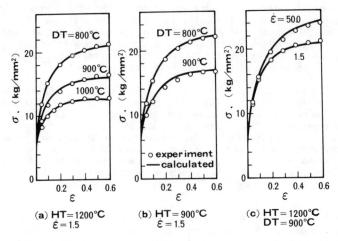

Figure 8–Effect of HT, DT and $\dot{\varepsilon}$ on stress strain curves

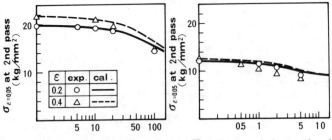

(a) HT $= 950°C$, DT $= 800°C$
    $\dot{\varepsilon} = 1.5$

(b) HT $= 1200°C$, DT $= 1000°C$
    $\dot{\varepsilon} = 1.5$

Figure 9–Decrease in deformation stress at the 2nd pass due to recovery of γ

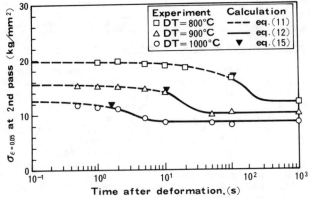

Figure 10–Decrease in deformation stress due to recovery and recrystallization of γ

tion period. Fig. 10 shows the decrease in the stress due to recovery and
recrystallization according to the time after deformation. The progress of
both recovery and recrystallization was calculated in this case. Lines and
solid triangles represent calculated results with eq.(11), (12) and (15).
Both experimental data (open marks) and calculated results are relatively in
good agreement, which implies that this mathematical model is available to
estimate the recovery and recrystallization behaviors in plate rolling process.

Discussion

Effect of Nb on deformation stress

The decrease in observed stress between the 1st and the 2nd pass comes
from the difference between the softening due to recovery and recrystallization and the precipitation hardening of Nb, but they cannot be analysed
separately because they interact each other. Fig. 11 shows the effect of Nb
content on the stress. The stresses of the specimens held for 500s at 800°C
for the precipitation before deformation were almost the same as the stresses
of the specimens without holding. From Fig. 11, the increase in the stress
is expected to be proportional to the Nb content. As the specimen in the
present study contains only 0.01% Nb, the precipitation hardening is expected
to be no more than 1 kg/mm^2 which is within the error of observed stress.

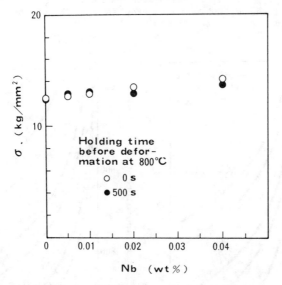

Figure 11-Effect of Nb content on deformation stress

Precipitation behavior of NbCN before deformation might be different
from that after deformation.[22] In this respect, DeArdo[23] reported that NbCN
selectively precipitates at grain boundary and sub-grain boundary of γ.
Therefore the effect of these precipitates on deformation stress might be
negligibly smaller than that of coherently precipitated NbCN at matrix.

Though the precipitation hardening of Nb is neglected in the present
model, both experimental data and calculation results show a good agreement.

Critical condition of recrystallization after deformation.

Incubation period, τ is calculated as a function of ε and T in eq.(15). The effect of deformation conditions on the critical conditions of recrystallization can be analyzed by eq.(15). Fig. 12 shows the effect of ε on τ. In the case of multi deformation test, ε in abscissa is calculated with eq. (8). Calculated results indicate that the higher deformation temperature and the smaller γ grain size result in the smaller incubation period. Fig. 13 shows the critical conditions of recrystallization of γ. The right upper sides of the lines are the recrystallization ranges and the left lower sides are the no recrystallization ranges. If interval time during deformation is longer than incubation period (parameter in Fig. 13), recrystallization starts during interval time. Therefore short interval time extends the no recrystallization range to the higher temperature side. Fig. 14 shows the effect of ε on the softening ratio (S) of double deformation test just at the start of recrystallization. S is described as[4]

$$S = (\sigma_m - \sigma_n) / (\sigma_m - \sigma_0) \tag{16}$$

where, σ_0 ; yield stress of the 1st pass
σ_m ; peak stress of the 1st pass
σ_n ; yield stress of the 2nd pass

Stresses, σ_0, σ_m and σ_n is calculated from eq.(2), (8), (11) and (15). In the calculation, σ_0 and σ_n were taken as $\sigma_\varepsilon = 0.05$ for the 1st and 2nd pass, respectively.

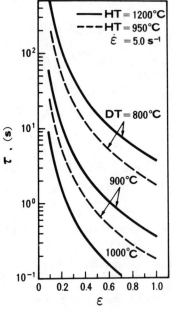

Figure 12-Effect of strain on incubation period

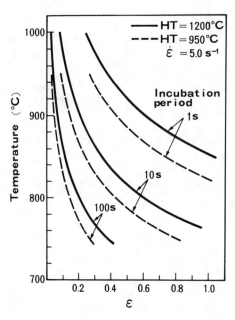

Figure 13-Critical conditions of recrystallization of γ

152

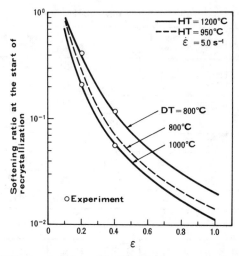

Figure 14-Effect of deformation conditions on softening ratio at
the start of recrystallization

Some authors [4],[5] reported that static recrystallization started at the specific value of S; the value of S, however, changes widely with the variation of deformation conditions in Fig. 14. The criterion of recrystallization is not necessary to be described as a specific value of S in general conditions. On the other hand the present model makes it possible to calculate the critical conditions of recrystallization in any deformation conditions of the hot rolling of plates.

Application of the present model to the rolling process

The present mathematical model has been applied to the practical multi pass rolling and the relation between the calculated retained strain due to controlled rolling and the mechanical properties of the plates have been investigated. The steel B was used for rolling. The slabs of 60mm thickness were heated to 1050°C and rolled to 20mm thickness with various conditions shown in Table V followed by water cooling at the cooling rate of 30°C/s.

Fig. 15 shows the changes of calculated retained strain in rolling process of the specimens F3, F7 and F9. The solid marks represent the retained strain just before the water cooling (ε_f) which are almost the same as the retained strain just before γ/a transformation. Recrystallization did not start in the rolling of F3 and F9. In spite that F9 was rolled in the higher rolling temperature and the smaller strain per pass than those of F3, ε_f of F9 was larger than that of F3. This implies that the shorter interval times between passes prevented the decrease in retained strain of F9 due to recovery. In the rolling of F7 γ recrystallized in the first three passes and ε_f was very small. Fig. 15 indicates that the interval times as well as the rolling temperature strongly affect the retained strain.

Fig. 16 shows the relation between ε_f and the ductile-brittle transition temperature of the plates obtained by Charpy impact tests. Fig. 16 reveals that the increase in ε_f improves the toughness of the plate. It is well known that the deformation of γ makes the microstructure after γ/a transformation fine by increasing the nucleation rate and the nucleation site. The increase in ε_f calculated from average dislocation density by the

153

Table V. Rolling conditions

No.	T_S (°C)	T_F (°C)	No.of pass	t_i (s)	t_t (s)	average ε	average $\dot{\varepsilon}$ (s^{-1})
F 1	885	735	15	15.8	18	0.067	4.8
F 2	820	746	8	13.3	43	0.115	4.8
F 3	780	765	5	13.8	21	0.202	4.8
F 4	900	732	15	12.6	17	0.067	2.8
F 5	830	751	8	12.7	20	0.115	2.8
F 6	820	754	8	12.1	35	0.115	4.8
F 7	900	763	5	40.0	23	0.202	4.8
F 8	900	733	8	25.3	28	0.115	2.8
F 9	800	776	8	4.6 (1.9, 8.3)	5	0.115	7.3
F 10	780	768	5	5.3 (2.2, 8.3)	5	0.202	7.3

T_S : Start rolling temperature
T_F : Finish rolling temperature
t_i : Average interval time
t_t : Transfer time from the finish of roll-
ing to the start of cooling

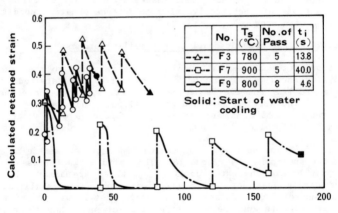

Figure 15-Change of retained strain in hot rolling process

present model is expected to promote the nucleation both by the increase in
the nucleation rate due to the change of the grain boundary structure and by
the increase in the nucleation site inside γ grain such as dislocations and
deformed structures.[24] The toughness of the plate, consequently, is improved
by fining of the microstructure through the increase in ε_f. As all the rolling conditions such as temperature, strain and interval time are taken into
consideration in the calculation of ε_f, it is possible to evaluate the effect
of the total rolling process on the toughness of the plate in a given cooling
conditions after the rolling. As seen in Fig. 15, in order to increase ε_f for

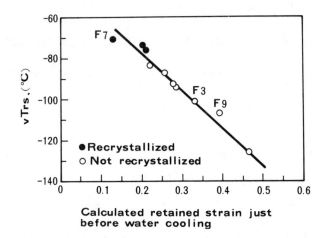

Figure 16-Relation between retained strain and vTrs

the purpose of improving the toughness of the plates by the control of the progress of recovery and recrystallization, it is important to to keep the rolling temperature in lower temperature range and to finish the rolling in shorter period.

Conclusions

The behaviors of static recovery, recrystallization and the retained strain of γ in plate rolling process have been formulated in terms of the change of average dislocation density calculated from the decrease in deformation stress observed by the double deformation tests. Furthermore, the relation between the calculated retained strain due to controlled rolling and the toughness of the plates have been investigated. The results are summarized as follows:
1. The relation among stress, strain and dislocation density and the change of dislocation density due to strain hardening and dynamic recovery during deformation have been formulated. By the present formulation, deformation stresses of various deformation conditions can be calculated as a function of temperature, strain, strain rate and γ grain size.
2. The behaviors of static recovery and recrystallization during holding period after deformation have also been formulated as a function of dislocation density. This formulation makes it possible to estimate the critical condition of recrystallization during the interval time between the successive rolling passes of plate.
3. Deformation conditions such as deformation temperature, deformation strain, strain rate, γ grain size and interval time during deformation affect the critical condition of recrystallization. The smaller interval time extends the no recrystallization temperature range to the higher temperature side.
4. Interval times between successive rolling passes as well as the rolling temperature strongly affect the retained strain in γ. The increase in retained strain just before γ/a transformation improves the toughness of the plate.

Acknowledgement

Authors wish to thank Dr. H. Sekine, Dr. H. Mimura and Dr. T. Senuma for many helpful discussion.

References

1. H. Sekine: <u>Proceedings of the 86th Nishiyama memorial lecture</u>, (1982), 123, ISIJ

2. T. Tanaka, N. Tabata, T. Hatomura and C. Shiga: <u>Microalloying 75</u>, (1977), 88, Union Carbide

3. Kozasu, C. Ouchi, T. Sampei and T. Okita: ibid., 120

4. R.A.P. Djaic and J.J. Jonas: <u>JISI</u>, (1972), 256

5. C. Ouchi, T. Okita, T. Ichikawa and Y. Ueno: <u>Trans. ISIJ</u>, 20 (1980), 833

<u>6.</u> C.M. Sellars: <u>Metal. Soc.,</u> London, 3 (1979), 3

7. T. Senuma, H. Yada, G. Matsumura and T. Futamura: <u>Tetsu to Hagane</u>, 70 (1984), 2112

8. Y. Saito, M. Tanaka, M. Kimura, T. Sekine, K. Tsubota and T. Tanaka: <u>Kawasaki Steel Technical Report,</u> 9 (1984), 12

9. A. Yoshie, M. Morikawa, Y. Onoe and K. Itoh: <u>Trans. ISIJ,</u> to be published

10. T. Senuma, H. Yada, G. Matsumura, S. Hamauzu and K. Nakajima: <u>Tetsu to Hagane,</u> 70 (1984), 1392

11. Japan Institute of Metals: "<u>Ten-i-ron</u>", (1981), 66, (Maruzen)

12. H. Yoshinaga: <u>Proceedings of the symposium of numerical analysis on hot working of metal</u>, (1985), 1, (JSTP)

13. J.E. Bailey and P.B. Hirsch: <u>Phil. Mag.,</u> 5 (1960), 485

14. W.G. Johnston and J.J. Gilman: <u>J. Appl. Phys.,</u> 30 (1959), 129

15. R. Sandstrom: <u>Acta Met.,</u> 25 (1977), 897

16. S. Suzuki: "<u>Kinozoku no tuyosa</u>", (1981), 154, (AGNE)

17. J.W Cahn: <u>Acta Met.,</u> 4 (1956), 449

18. D.R. Barraclough and C.M. Sellars: <u>Met. Sci.,</u> 13 (1979), 257

19. T. Terazawa, A. Yoshie, Y. Onoe and K. Nakajima: <u>Tetsu to Hagane,</u> 69 (1983). S631

20. Y. Misaka and Y. Yoshimoto: <u>J. Japan Soc. for Tech. Plast.,</u> 8 (1967), 414

21. S. Lička, L. Zela, E. Piontek, M. Košař and T. Prnka: <u>Proceedings of the international conference on steel rolling,</u> (1980), 840, (ISIJ)

22. B. Dutta and C.M. Sellars: Private communication at the congress of the committee of process metallurgy in hot rolling of ISIJ, 19 Sept., (1985)

23. A.J. DeArdo, J.M. Ray and L. Mayer: <u>Proceedings of the international symposium of Niobium,</u> (1981), 685, (TMS)

24. M. Umemoto: <u>Proceedings of the symposium of simulation of hot working and transformation of austenite,</u> (1985), 17, (ISJI)

Quality Optimized Process Control

of Open Die Forging

E. Siemer, P. Nieschwitz and R. Kopp

Intitute of Metal Forming, RWTH Aachen
5100 Aachen, Intzestr. 10, Germany

Abstract

Economic production of high alloy steel forgings requires an optimization of the forging process in terms of productivity and quality. The investigations covered the flow of material both in the contact area between forging tool and workpiece and in the core zone during the draw out of square bars with flat tools. Edge and surface cracks may occur thereby in case of multiple-pass forging of materials difficult to deform, as a result of the superposing of local strain maxima in the tool radius range, which can amount to a multiple of the average elongation rate. Based on these experimental and theoretical researches, an online process control system was developed, which optimized the forging schedule according to exactly defined technology rules. The use of modern measuring and control equipment will provide that the increasing requirements concerning quality and flexibility of this forming process will be met.

Introduction

In the coming years, the application of flexible production systems will become increasingly important in forming plants. Computer controlled open die forging is an example. This is a forming process in which simple tool forms can be used to produce complex workpieces. By varying process parameters such as tool form, press stroke, bite ratio, tool velocity and possibly using supplementary heating or cooling equipment, it is possible to influence the local technological forming quantities: plastic strain ε_{ij} , strain rate $\dot{\varepsilon}_{ij}$, temperature ϑ and stress σ_{ij}; and with these, finally the material qualities, Fig.1.

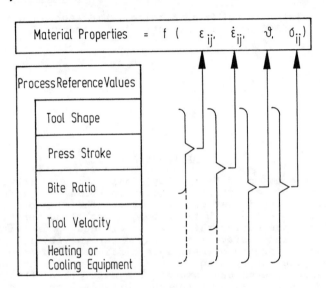

Figure 1. Technological means of controlling the open die forging process.

For experimental research on the forging process, a fully automatic forging machine is at the disposal of the institute of metal forming (IBF RWTH Aachen), equipped with expanded process control and measuring features.

Conception of measurement and control

The system configuration of the forging equipment is divided into three levels, Fig.2. A superior process computer with peripheral devices for operating the process software and a data registration system are included in the system. After determining the type of forming strategy and fixing a quality criterion, the forging program is generated with a process model. With the resulting NC-data the controlling computer calculates the nominal values for the process guidance. The measured actual values are partly fed back into conventional control circuits and with the help of the data registration prepared as a process log.

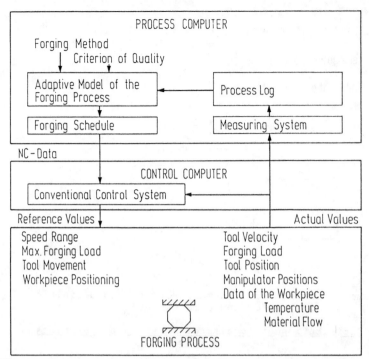

Figure 2.   Process control structure in open die forging.

A model of the forging process can naturally only describe the
process conditions in parts, and even that with inaccuracies.
Thus, adaptation means fitting the mathematical model by regis-
tering the deviations from nominal values, so that the forging
process will run within the allowed tolerances.

Criteria for the optimization of quality

and productivity

The basis has been established within the framework of a
research project promoted by the German Research Society
(DFG), firstly in forward calculations and optimization of the
forging process for simple geometries. The aim of these investigations was to analyze the draw-out process bite by bite from
the point of view of the outline and core area material flow as
well as displacements in the contact area tool/workpiece, required force and energy and local temperature distribution.
Fig.3 shows the principle of the draw-out process and defines
the characteristic parameters used here.

rel. reduction in height $\varepsilon_h = \left| \dfrac{h_1 - h_0}{h_0} \right|$ spread $\beta = \dfrac{b_1}{b_0}$

elongation rate $\lambda = \dfrac{l_1}{l_0}$

bite ratio s_b / h_0

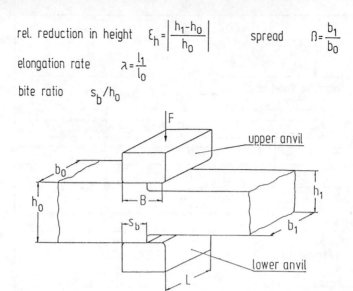

Figure 3. Principle of hammer forging and definition of the forging parameters.

Distribution of local strain maxima on the block surface

In bar steel forging of materials with low workability, flaws can be observed on the surfaces in form of edge cracks. Simulation of the draw-out process with two-dimensional FEM shows that distinctly greater strains exist in the tool edge regions than in the contact area tool/workpiece, Fig.4. These calculations were accomplished with the boundary condition "sticking

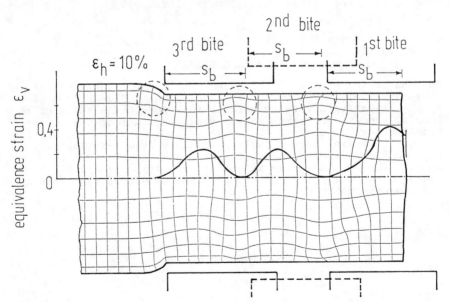

Figure 4. Distorted FE-net for a draw-out process /1/.

Experiments carried out in parallel to these calculations on bars with surface grids confirmed this assumption, Fig.5. Localized strain maxima $\lambda_{R(\ddot{o})}$ arise on the block surface in the tool radius range, while sticking friction exists in the other regions. These localized strain maxima are dependent on the reduction in height and the bite ratio.

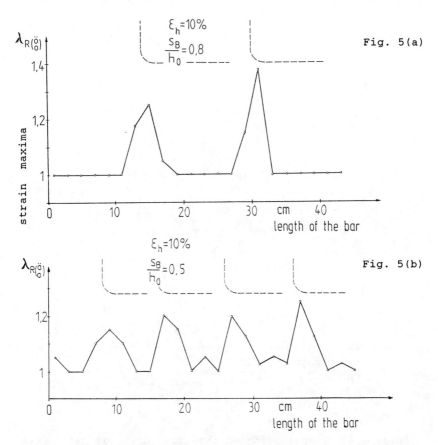

Figure 5.   Distribution of local draw-out rates in the contact area tool/ workpiece over the length of a forging bar for two different bite ratios.

It should be interesting to know what will happen when, for multiple passes, the tool radius is repeatedly positioned on the same block area. Will superposition of local strain lead to strain maxima, which caused the above described cracks?

In order to answer this question, the cross section of a forging was reduced step-wise by repeated passes without turning the bar. With the elongation of the bar taken into consideration, it was attempted to position the bite edge as close as possible to the bite edge of the preceding pass for each stroke, Fig. 6(a). By superimposing the local strains, elongation rate maxima of up to six times the avarage elongation rate were produced in the tool radius influence zones. On the

other hand, sticking friction was measured in the contact area
tool/workpiece. A second bar was forged with the same boundary
conditions but with bite offset, excluding the effect bite edge
on bite edge. Fig.6 bottom shows that peak levels are nearly
even in this forging process. Thus, superposition of local
strain maxima can be avoided in forging of ingots with bite
offset.

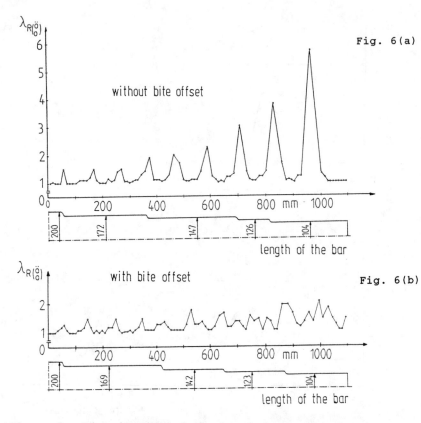

Figure 6.   Distribution of the local draw-out rates over
            the length of a stepped forging bar with and
            without bite offset.

To demonstrate the reproducible formation of cracks, a marginally deformable aluminum alloy AlMgSi1 was chosen as representative of materials of low formability. With this material,
defined edge and surface cracks could be produced at positions
determined beforehand, Figs. 7 and 8. These defects could be
avoided by employing exactly defined bite offset under the same
forging boundary conditions.

Figure 7. Occurrence of edge cracks caused by the superposing of local strain maxima in a marginally
deformable aluminum alloy.

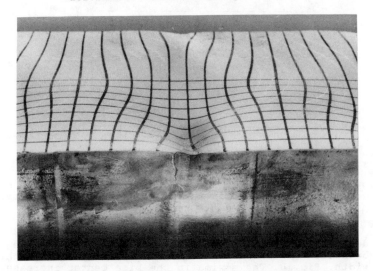

Figure 8. Detail enlargement of an edge crack.

Deformation pattern within the core zone

A further, important measure of the quality of a forging
product is the deformation of the core zone. Forging bars were
furnished with pins in the longitudinal axis for experimental
analysis, Fig.9.

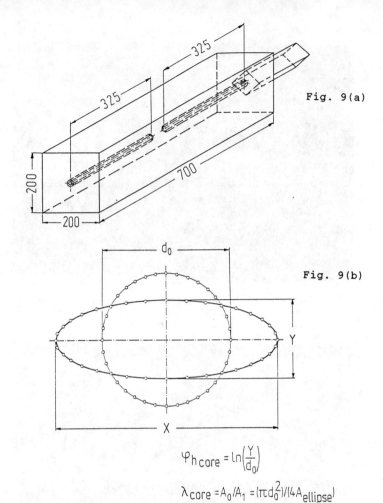

Fig. 9(a)

Fig. 9(b)

$$\varphi_{h\,core} = \ln\left(\frac{Y}{d_0}\right)$$

$$\lambda_{core} = A_0/A_1 = (\pi d_0^2)/(4A_{ellipse})$$

Figure 9.   Forging bar prepared for an investigation of the material flow in the core zone and definition of the measured values.

The deformation of the pin after one pass shows the very inhomogeneous distribution of the core penetration within a bite width, Fig.10. The maxima in the bite center increase with growing bite ratio and reduction in height.
These relations are illustrated in Fig. 11. Additional results of a 2-dim-FEM calculation are shown for comparison with the measured quantities. These correspond well for small height reductions or for small bite ratios, since the lateral spread can be neglected in these cases.

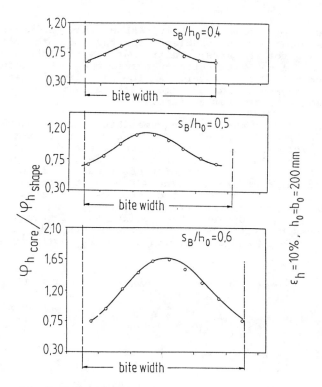

Figure 10. Deformation pattern in the core over the bite width for three different bite ratios.

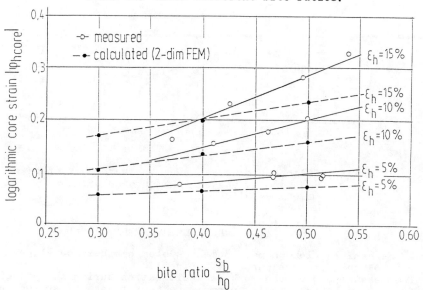

Figure 11. Comparison of measured and calculated deformations in the core as a function of bite ratio and reduction in heigth.

The search for optimum bite ratio shows that counter-rotating tendencies exist for the problematic nature of core deformation as well as for the strain distributions on the workpiece surface. A large bite ratio promotes the core deformation maxima under the bite center. At the same time it leads to a strong inhomogeneity in the core area and increases the local strain maxima on the bar surface.

Through-put, the number of intermediate heatings needed and energy required also play an important role in efficient production. Fig. 12 shows the influence of the bite ratio on the number of strokes and passes for a cross section reduction from □200 to □100. As a result of these studies, it may be said that a bite ratio of around sB/h0 = 0.5 can be considered as a good compromise for the described problems. Additional improvements in quality can be achieved by employing the forging strategy with bite offset. The realization of this technology is only possible with the help of modern computer controlled forging machines and the application of a length measurement system.

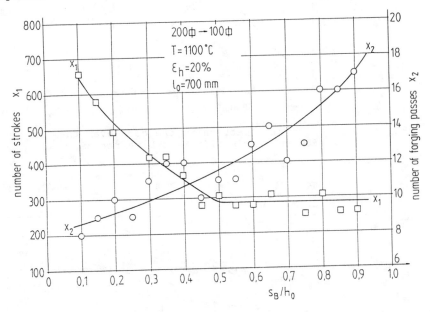

Figure 12. Effect of bite ratio on the number of strokes and forging passes in a forging process □200 ⟶ □100 .

Online length measurement

A mechanical length measurement system has been developed at the IBF for research purposes, Fig. 13.
The measurement of the workpiece elongation during the forging process (online) leads to empirical equations for spread and elongation for different materials and geometries. The forging of a carbon steel with a starting cross section of 200□ and an end cross section of 140 □ serves as an example. The forging

schedule is calculated using a spread and elongation equation
of Tomlinson/Stringer /2/, Fig. 14.

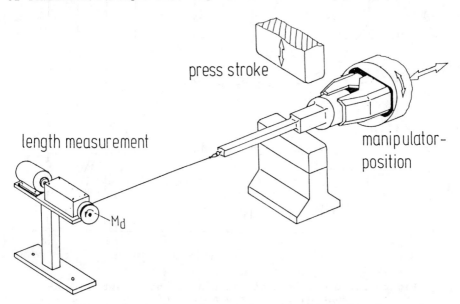

press stroke

length measurement

manipulator-
position

M_d

Figure 13. Mechanical equipment for measuring the elongation
of the workpiece during the forging process.

STICHPLANBERECHNUNG QUADRAT   forging schedule

EINGABE:

BEZOG. HOEHENAENDERUNG EH [%]:   10.0   reduction in height

BISSVERHAELTNIS BH=SB/H0 ...:   0.5   bite ratio

ANFANGSLAENGE   LA [mm] ....:   500.0   initial length

ANFANGSHOEHE   HA [mm] ....:   200.0   initial heigth

ENDHOEHE   HE [mm] ....:   140.0   final heigth

STNR	WINK grd	EH %	SB mm	H0 mm	H1 mm	B0 mm	B1 mm	L1 mm
1	0.	10.0	100.0	200.0	180.0	200.0	206.1	539.1
2	90.	9.8	103.0	206.1	185.9	180.0	185.9	578.9
3	0.	10.0	92.9	185.9	167.3	185.9	191.5	624.3
4	90.	9.8	95.8	191.5	172.7	167.3	172.7	670.3
5	0.	10.0	86.4	172.7	155.5	172.7	178.0	722.8
6	90.	9.8	89.0	178.0	160.5	155.5	160.5	776.1
7	0.	10.0	80.3	160.5	144.5	160.5	165.4	836.9
8	90.	9.8	82.7	165.4	149.2	144.5	149.2	898.6
9	0.	6.2	74.6	149.2	140.0	149.2	151.9	940.4
10	90.	7.8	76.0	151.9	140.0	140.0	143.5	995.7
11	90.	2.4	71.7	143.5	140.0	140.0	141.0	1013.2

Figure 14. Example of a precalculated forging schedule
for a reduction □200 ──▶□140.

Confined to this material and the chosen square cross section
this calcultion corresponds well with the experiment, Fig.15.

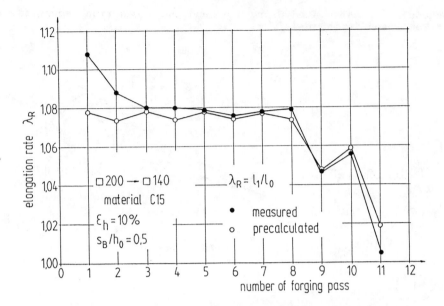

Figure 15. Comparison of calculated and measured elongation rate for a number of forging passes.

Other materials will produce deviations in spread and elongation. Influences like different bar-shapes will also lead to a
non-sufficient precision of predicted schedules. For controlled
forging the process model must be adapted to the actual process
with the help of the length measurement in order to allow the
exact positioning of the workpiece. Such process models are
being developed and tested at the IBF.
The discussed mechanical length measurement system is surely
not adequate for the rough conditions of a forging plant.
Considering the rapid development of non-contact length measurement systems on the basis of lasers or photodiode arrays,
it can be expected that such sensors will be available for
forging in the near future.

Process log

A further advantage of the use of a computer in open die forging is that a process-accompanying log of each forging pass
can be made with online registration. With the help of empirical and theoretical models, statements can be made concerning
the course of the forging process, and resulting statements on
the quality of the workpiece such as core penetration along the
centerline, Fig. 16. Thus, this process log can be seen as an
empirical-theoretical process accompanying quality control.

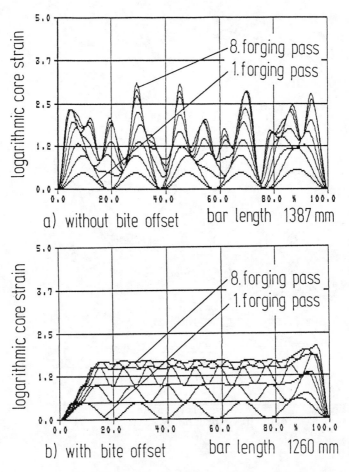

Figure 16. Representation of the equivalent core strain over the workpiece centerline after 8 forging passes.

Outlook

Fig.17 shows the information flow of a future open die forging plant. During operation scheduling, optimized forging programs are made with the help of simulation programs. The resulting NC-data are then used for the automatic production of forgings. Process models with sensor feedback ensure the compliance with geometrical tolerances and quality requirements. The quality control will have a prepared forging log at the disposal in order to control the forging products. Deviations from the quality requirements are fed back into the planning of the operations scheduling. The connection of the different areas by the use of a central computer ensures a smooth and controllable data flow avoiding time consuming and faulty interfaces between the different system parts.

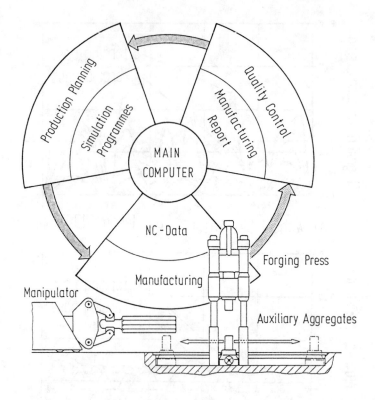

Figure 17. Information flow in an modern
open die forging plant.

References

1   Kopp,R., Nieschwitz,P., Cho, M.L.:
DFG-Forschungbericht KO579/12, RWTH Aachen, 1986

2   Tomlinson,A., Stringer,J.D.:
J. Iron Steel Inst. 193 (1959) 2, P. 157/162

Process Control

ABSTRACT

"THE INTELLIGENT PROCESSING OF ADVANCED MATERIALS"

Dr. Brian G. Kushner
The BDM Corporation
7915 Jones Branch Dr.
McLean, VA, 22102

and

Dr. Phillip A. Parrish
Defense Advanced Research Projects Agency (DARPA)
Defense Sciences Office
Materials Sciences Division
1400 Wilson Blvd.
Arlington, VA, 22203

The Intelligent Processing of Materials (IPM) program is an interdisciplinary research thrust initiated by DARPA during the past year. It seeks to combine the expertise of the laboratory materials scientist and of the process engineer, through the use of expert systems, with in-situ sensors and process models to facilitate greater control over the materials processing environment. The program began in response to the recognized need within the materials processing community for advanced techniques to improve the quality, reliability and yield of new materials. As a result, the program seeks to reduce the long lead time between initial materials development and eventual materials production and utilization. This paper will provide an overview of this program, the key technologies being developed and major thrust areas for the future.

"THE INTELLIGENT PROCESSING OF ADVANCED MATERIALS"

Dr. Brian G. Kushner
The BDM Corporation
7915 Jones Branch Dr.
McLean, VA, 22102

and

Dr. Phillip A. Parrish
Defense Advanced Research Projects Agency (DARPA)
Defense Sciences Office
Materials Sciences Division
1400 Wilson Blvd.
Arlington, VA, 22203

INTRODUCTION

The Intelligent Processing of Materials (IPM) program is an interdisciplinary research thrust initiated by DARPA during the past year. It seeks to combine the expertise of the laboratory materials scientist and of the process engineer, through the use of expert systems, with in-situ sensors and process models to facilitate greater control over the materials processing environment. The program began in response to the recognized need within the materials processing community for advanced techniques to improve the quality, reliability and yield of new materials. As a result, the program seeks to reduce the long lead time between initial materials development and eventual materials production and utilization.

The use of artificial intelligence in materials processing, in the form of expert systems, is a radically different application of this new and vital technology. Previously, expert systems have been confined to the research lab and as assistants a variety of analysts. In each case, however, the domain of expertise was limited in scope and in performance, typically functioning at rates of 10 - 100 RIPS (**Rule Inferences Per Second**, a measure of performance in expert systems). Recent advances in AI, and expert systems in particular, have, for the first time, made the application of this technology to materials process control within reach.

The use of expert systems in the materials processing environment, in addition to being both data and knowledge intensive, requires that the system reach conclusions within the decision timeframe of the process under control. For the current generation of expert systems technology, this translates into timescales for active process control on the order of minutes to tens of minutes. While at first glance this may appear to be a fairly stringent requirement, upon further analysis it is evident that many materials systems satisfy this criteria. In particular, a single step process like the growth of bulk crystals from a solution or melt, as in the case of semiconductors such as Gallium Arsenide (GaAs), requires process decision in timeframes of minutes. The situation is very similar for a multi-step process such as the fabrication of composite materials, like carbon/carbon composites. While the near term focus of the IPM program will be on these two materials systems, additional efforts will within the program will focus on identifying opportunities for development of other IPM-based materials processing systems and on the techniques to achieve rapid prototyping of IPM knowledge bases to overcome existing bottlenecks in the IPM programs.

The use of expert systems to directly control the quality of the material produced, as opposed to the quantity, necessitates the coupling of two different types of expertise. On the one hand, the laboratory materials scientist has detailed knowledge about the materials themselves, and possesses a laboratory which is instrumented with numerous sensors and detection devices. This knowledge is being applied only to a small quantity of processed material, and many decisions can in fact be made "off-line." In the production environment, off-line interpretation is very costly, and the process engineers must make corrective decisions in semi-real time. Furthermore, most materials production facilities are not as heavily instrumented as research labs. This requires that the process engineer infer the properties and characteristics of the batch process from either small samples removed from the process or via periodic measurements of key process parameters. In either case, the process engineer bases decisions upon prior experience rather than detailed knowledge of the materials and process kinetics. A major goal of the IPM program is to integrate these two types of materials processing expertise within a tool that can facilitate improved production materials quality.

By developing a set of tools and an intelligent controller to aid materials production, DARPA seeks to revolutionize the way in which materials developments are transitioned from the lab into production. With expert systems as the catalyst, it is hoped that the IPM program will provide a framework for further incorporation of artificial intelligence in materials processing. Thus, within the same program, there are efforts to: advance the processing of cutting edge materials; develop powerful new process models capable of predicting process control envelopes for complex materials; implement a set of real-time, in-situ sensors to provide direct process variable information to the intelligent control system; streamline the knowledge acquisition process in the development of intelligent controllers; and analyze the structure-function relationships between devices and materials. The end goal of the program is to drastically reduce the time associated with the utilization of novel materials in advanced systems.

This paper will address the central aspects of the IPM program, the key technologies being investigated and the expected impact on the materials processing community. Since the IPM program is an extremely broad effort, conducting research in both the role of expert systems in materials manufacturing and in optimizing the underlying materials properties that one seeks to control through the use of intelligent controllers. This remainder of this article is divided into four sections focusing on IPM research in GaAs bulk crystal growth (section II), on carbon/carbon materials processing (section III), on knowledge acquisition tool research (section IV) and on directions for future program activity (section V).

II. GaAs Liquid Encapsulated Czochralski (LEC) Bulk Crystal Growth

The IPM effort in GaAs addresses the problems associated with the High Pressure Liquid Encapsulated Czochralski (HP-LEC) growth of bulk Crystals. The basic objective of this research is the design and implementation of an integrated IPM crystal growth system which will maximize yield and minimize human interaction for the economic production of large area (> 4 inch diameter), long (6 - 10 inches), high yield, high quality, high uniformity HP-LEC GaAs. The prime contractor on this project is the General Electric Co, which is teamed with scientists from Spectrum Technology, Inc (GaAs production) and Nektonics, Inc. (process models) to demonstrate pilot production of a GaAs IPM system. This project was initiated in late 1986, and is the most evolved of the IPM technology development programs.[1]

In an effort to maximally integrate IPM concepts with existing puller technology, baseline crystal growth experiments are being conducted in conjunction with a series of modifications to a Cambridge Instruments computer controlled puller. These tests are being used to define furnace conditions and process control parameters for uniform, low-impurity, twin-free growth with specified electronic properties. The puller itself is being modified to incorporate sensor access ports and is being outfitted with a digital controller to enable rapid database generation and historical trends analysis.

The sensor techniques being developed will focus on measurement of direct process control variables as inputs to the process models and knowledge based controller. The critical problem is the determination of the relationship between directly defined crystal and melt morphologies (e.g., as defined by x-ray) and indirect properties from thermophysical (e.g., as defined by IR imagery) and visual observations. The x-ray inspection module has been successfully employed to make computer tomographic images of the cross sections of growing crystals. This is furnishing critical information about the melt interface as a function of different melt depths, varying crystal and crucible rotation speeds and twinning/twin-free growth conditions. These scans are being processed with the aid of GE's Image Processing Development Environment.

The application of IR imaging techniques to the HP-LEC process is not straightforward due to the present of the boric oxide encapsulant. Direct usage of thermal

IR imaging radiometers is limited by the need to spectrally filter inputs to respond to wavelengths where the B_2O_3 emittance approaches unity. Utilization of near IR radiometers will require calibration for effects of varying B_2O_3 emissivity. The radiative properties of the boric oxide, as well as that of the solid and liquid GaAs, will be determined in-situ, allowing for the creation of a more reliable database of these parameters for use in IR image interpretation and process modeling.

Using these IR sensors, efforts will be directed at obtaining infrared images of: the exposed portion of the boule, the upper and lower surfaces of the boric oxide, the meniscus and the melt interface on the underside of the boule. This profile imagery will be correlated with the computer tomographic x-ray cross sections to validate and quantify these measurements. Vision processing will also be applied to determine the instantaneous boule diameter, the presence or absence of twinning during growth and other surface features. Existing sets of fast conversion algorithms will be applied in the IR imagery interpretation process, allowing potential implementation of these sensors as real-time monitors of the thermophysical evolution of the HP-LEC process. The vision system, on the other hand, can be hard-wired for real-time operation once a suitable set of feature extraction and interpretation algorithms are developed.

Existing heat transfer and growth models for GaAs are being enhanced and implemented within a single model. While the task of developing an integrated heat transfer and melt convection model is beyond the scope of the present activity, decoupled calculations are still possible and are being pursued. The GaAs process models assume an axially-symmetric geometry for the crucible, furnace and boule, and seek to determine the shape of the meniscus and solid-liquid interface, the thermal and stress distributions in the growing boule, the temperature distributions at sensor locations and the details of the flow near the crystal. Numerical simulations are underway to account for the effects of boule/crucible rotation direction and rate, axial thermal gradients along the crucible wall, magnetic field strength and other flow control techniques within the furnace on the thermophysics and residual stress effects within the growing crystal. While these models may be exceedingly complex at first, a goal of this effort is to develop a reduced set of process control models capable of existing in a real-time PC/AT delivery environment.

The knowledge base of the IPM system is being developed within the Knowledge Engineering Environment (KEE, version 3.0) from IntelliCorp on a LISP machine. A prototype model of the IPM process is being created with process information hierarchically instantiated within the KEEworld system. Process descriptions and sensors are represented as frames, with diagnostics as procedural entities operating on the process frames, and process planning and causal/temporal relationships as frames in multiple KEEworlds. Modifications or corrections to the control strategy, as would be required in response to a deviation from the ideal production path, will be implemented as methods fired by changes in active values attached to the slots within process or sensor frames.

The knowledge system will be developed to support real–time GaAs crystal growth diagnosis and control. However, owing to the concurrent development of the related process models and sensors, the behavior of these IPM system components will be simulated to allow knowledge base development, verification and validation to proceed independently. The integration of these components and the hierarchical control system will occur late in the program. A VAX 11/780 is being used for model development and control algorithm design, while a MicroVAX II workstation is the target environment for the production implementation of the knowledge system. The PC–AT delivery environment will be linked via ethernet to the MicroVAX, creating a process monitoring and diagnostic control environment that exploits the best of model–based signal processing and AI system techniques. These capabilities, along with automated corrective action modules for detected growth failures, will be demonstrated in a pilot–production demonstration at the conclusion of this effort in late 1988. Successful demonstration of pilot production would pave the way for full–scale production of wider (greater than 4–inch diameter), longer (greater than 6–inch length), higher purity (less than $10^{13}/cc^3$) and uniformity (greater than 90%), improved yield GaAs HP–LEC bulk crystal growth by 1991.

III. Carbon/Carbon Processing

The motivation for pursuing an IPM program in carbon/carbon (C/C) production is that it is a multi–step fabrication process with relatively long process kinetics decision timeframes for which IPM–based improvements could have a significant impact. It was estimated that IPM–type process enhancements could lead to factors of 10^2 to 10^3 improvements in the performance, reliability and quality of C/C composite materials. As a result, an effort has been initiated in C/C that paralleled developments in the IPM growth of GaAs LEC bulk crystal growth (LEC–BCG), but included some modifications to account for the lower maturity level of the C/C domain. The decision was also made to concentrate activities on a critical subunit in the C/C process: carbonization. This subunit was chosen for the following reasons: it is central to the control of the end product, dictating density, yield and uniformity in carbon materials; it is the costly and critical lengthy step in C/C processing; and the lack of adequate sensors leads to heuristic intensive decision processes. Furthermore, the thermal evolution of the composite from early cure through late carbonization and subsequent cool–down requires multivariate control over chemical, microstructural and thermophysical parameters to insure mechanical integrity in the end product. As such, it is a challenging domain for the exercise of IPM technology. Finally, it was also apparent that success of the IPM concept for this area within a two year timeframe would enable rapid transition to other subunits, allowing a full IPM C/C process controller to be developed.

Carbon/carbon is a particularly challenging domain because, at present, the ability to provide real-time, in-situ monitoring and interpretation of the carbonization and graphitization processes is at the leading of research edge at the present time. In focusing on the carbonization unit process, an initial two year period would develop I/O relationships between sensor interpretation algorithms and process model predictions. This will serve as a prelude to development of a pilot scale intelligent controller. As presently structured, the IPM carbon/carbon composites effort requires research to be conducted in four areas, with Rockwell International as the prime contractor (models, sensors & controller development), with assistance from Kaiser Aerotech (C/C production knowledge) and GA Technologies (composite process knowledge).

The first areas of focus are the development of real time in-situ sensors which are critical to the carbonization process and the development of techniques for the unambiguous real-time interpretation of the sensor data. This includes sensors which are required to obtain both a macroscopic and microstructural understanding of the c/c carbonization process. Thermocouples will be used to provide the requisite component temperature information, since the rages of interest (approximately $3000C$ to $10000C$) are within the capabilities of accurate thermal measurement and interpretation. Gas chemistry analysis will be applied throughout the carbonization process, drawing on the experiences and techniques developed in the fossil fuel industries for monitoring carbon and hydrocarbon products. A suite of gas chromatographs are particularly well suited for identifying and measuring the volume and composition of volatile pyrolysis reaction products. Gas measurement sensors will monitor the rate of evolution of H_2, CO, CO_2, CH_4, water and other hydrocarbons to determine reaction rates, reaction stages and the nature of the bonds being formed as a function of time.

Acoustic emission (AE) techniques will be applied in-situ to the evolving C/C product to identify stress defect developments such as delamination or crack formation. A challenge in the application of AE will be differentiating the acoustic signature of low level stress relief during carbonization from that due to microstructural damage. Microwave reflectivity will be used to monitor the degree of carbonization during the reaction arising from changes in the complex permittivity of the C/C sample. The influence of graphitic reinforcement on the reflectance properties of the sample in the microwave region will be explored in greater detail, as will effects due to dimensional (principally sample thickness variations) changes during the evolution of the carbonization process.

As in the case of GaAs, specification of the relationship between sensor outputs (from post-signal processing interpretation) and the set of required inputs for any process models is a major challenge in this IPM program. The emphasis will therefore be on the development of process models of C/C composite fabrication which are critical to the rapid, reliable production of C/C, focusing on control and understanding of the matrix microstructure. These models, of which the APIC (Analytical Processing for Improved

Composites)2 is the best developed, will emphasize the control of matrix microstructure to improve final composite properties and defect minimization. As part of this effort, extensions of this and other process models to 3-D will be investigated.

To insure mechanical integrity at the end of carbonization, the process models must account for fracture (the principal catastrophic failure mode) involving both fiber breakage and delamination or via crack-induced property degradation. Delamination in 2-D parts will be controlled through development of a local mechanical integrity model (LMIM). LMIM will draw upon temperature and volumetric gas analysis to estimate the current internal pressure during carbonization and correlate it with an allowable maximum internal pressure calculated from off-line finite element model stress analysis and AE data. These outputs will be fed into an intelligent controller that will track historical trends during carbonization and will maintain a margin between the maximum allowable internal pressure and that inferred from sensor inputs. When integrated with a knowledge base for historical trends analysis, process prediction and scheduling, this controller will be able to diagnose and correct aberrant process behavior prior to catastrophic failure.

The demonstration of an integrated IPM architecture that effectively integrates this knowledge-based controller, the real-time, in-situ sensors and the selected process models for carbonization process control is expected in late 1990 or early 1991. Demonstration of the control of materials quality, reliability processing and yield in pilot c/c production implementation of IPM controller for selected phases of c/c process could then occur by a scale-up of the carbonization IPM system to a multi-step controller. The overall goal of this effort is the development of a robust IPM system capable of comparable functionality using different sets of C/C composite manufacturing expertise.

IV. Knowledge Acquisition Tools

As a result of reviewing the roles of knowledge engineer and expert in the IPM program, and, more generally, in expert systems development, the knowledge acquisition bottleneck was identified as the single largest barrier to success in the IPM program.3 The development of effective knowledge base prototyping tools could combat this problem. This would enable the knowledge engineer to focus on his/her expertise (programming) and the expert to accentuate knowledge articulation. Central to this tool are facilities that promote productive engagement with the expert, overcoming the variety of problems associated with the expression of knowledge by experts. These knowledge acquisition facilities are being prototyped by The BDM Corporation for eventual incorporation into advanced IPM systems.

Experts are generally capable of bringing to bear a depth of understanding exceeding that of documented reference works. The expertise arises from a strong base of fundamental knowledge, problem solving approaches and tactics. This base allows the expert to function optimally in recognizing inconsistencies or errors in presented data. It is this same depth of understanding, however, which inhibits experts from extemporaneously summarizing their knowledge in a readily transferable format.

The typical approach to building knowledge bases for AI systems entails prolonged and intensive one–on–one interactions between a programmer and a domain expert. In this setting the programmer may elicit large amounts of information from the expert, impose his/her own organization onto this information, and encode it into the system. To date, this approach has forced expert system development to occur in the following two modes: either the knowledge engineer must develop an appreciation of the subtleties of the field under consideration or the expert must learn the workings and structure of the knowledge base and interact with it directly. Thus the knowledge acquisition process becomes one of knowledge "impedance matching," where neither the expert nor the knowledge engineer are utilizing the interaction time optimally. Simultaneously, there is the "knowledge conversion" problem: taking the knowledge obtained from domain experts in a structured fashion and converting it into a form that can be readily accessed by the expert system itself. While there have been many recent advances in expert system building technology such as expert system shells, but these advances do not address either of these bottlenecks to knowledge acquisition.

An experimental knowledge acquisition system, known as the IPM knowledge acquisition tool or KAT, has focused on the investigation of remedies for the knowledge engineering problem. KAT is being designed for use by a programmer as a tool to aid in the knowledge acquisition process, providing structured mechanisms for eliciting information about materials processing. In light of the above difficulties, it is important to realize that although KAT can assist in the automation of the knowledge engineering process to some degree, it is not envisioned as a replacement for the knowledge engineer. Instead it is best thought of as a tool that the knowledge engineer uses to query the expert in a structured manner. Such a tool in itself will constitute a marked advance in the knowledge engineering process.

Using KAT, the knowledge engineer and domain expert will iterate through three querying modes in which the expert will be prompted to answer different types of questions about a particular materials processing domain. In the Clarification Mode, the expert will be provided with facilities for creating an AND/OR graphical representation of the process. Once a certain amount of information is encoded in this format, the system will switch into a Prediction Mode. At the start of the Prediction Mode, the knowledge structure created during the Clarification Mode will be converted into a flowchart display. The user will be prompted to answer a series of questions about each subprocess represented in that structure. For instance, the expert might be asked to estimate ranges for temperature, pressure, and exhaust gas ratios at each point identified so far for the process. In addition, the expert will be asked to make predictions about what could go wrong at each given point, and to provide suggestions for redesign that might lead to avoidance of identified problems. Once these queries are completed, the knowledge engineer will return to the Clarification Mode, and recycle through the questioning, adding to the knowledge base created thus far.

Once the expert has iterated through the first two modes several times, a fairly detailed knowledge structure will be in place. At this point, the Diagnosis Mode will be invoked. In this mode, KAT's facilities will provide a step–through presentation of the materials process as specified up to this point. Sensor readings will be displayed so that the user can see how they change dynamically as the process progresses. In addition, the user will be able to trace the pattern of rules that are being evoked to affect these changes along the way. In this environment, then, the expert will have opportunities to diagnose and adapt the knowledge base created thus far, adding and deleting elements as necessary. Facilities to support this adaptation and modification of the knowledge base will be provided within the KAT system.[4]

Although the design of KAT is general enough to be implemented in any materials processing domain, the initial efforts will draw on knowledge in the area of Chemical Vapor Infiltration (CVI). Prototyping concepts for KAT have utilized the Knowledge Engineering Environment (KEE) 3.0 and SIMKIT environments on two Symbolics 3670 machines. KAT is drawing upon many of the facilities that KEE provides to aid AI programmers in the generation of frame–like objects and rules, as well as the graphics and specialized mouse functions which are useful in constructing user interfaces. The SIMKIT system, which resides on top of the KEE kernel, is being used for writing event–driven simulations and generating an overall model of the CVI process. Before a process model can be built in SIMKIT, a library must first be constructed which contains all objects and rule classes that will be used in the final system. Such objects include reactor types, preform types, information about thermal and pressure gradients, and so on. For each of these objects, further information may be stored in internal "slots". For instance, the object corresponding to a one–dimensional fibrous preform might have slots for such features as size, material makeup, initial density, optimum density, and reactive gas, to name a few. Each of these slots is allowed to take on certain values, and can be set and reset for any particular materials design scenario. The integration and demonstration of the prototype KAT system is expected in late 1989, making it available in time for utilization in the next generation of IPM program activities.

V. IPM and the Future

In the next few years, the IPM concept will be extended to achieve intelligent, low cost, automated manufacturing of advanced materials. As such, the IPM program seeks to accelerate the transfer, and hence reduce the overall time to deployment, of materials science developments in the laboratory to the production environment. This will be accomplished by combining the three traditional IPM components with knowledge based systems for concurrent product design, development and scheduling, rapid prototyping of low volume systems and the creation of high level specification languages to enable functional materials specifications in a prototype facility termed a "microfactory." A microfactory is what the name implies: a small, prototype and prototyping factory which produces a sufficient quantity of material output to validate full-scale production scaling. Increased quantities of output product are possible via replication of the microfactory. As such, a microfactory is the intermediate state between laboratory or benchtop materials experimentation and large-scale production and serves as a vehicle for achieving the overall goal of reducing defense materials and systems life cycle times (from R & D to production) from over 20 years to between 5 - 10 years.

The IPM program is currently investigating materials systems which can demonstrate this rapid transition from laboratory to microfactory to full-scale production. As an example, an emerging materials process known as rapid solidification by plasma deposition (RSPD) serves as a useful experimental base for testing the transfer of IPM techniques from laboratory to production. At present, RSPD exists only as a laboratory process, with limited computer control, few sensors, and poor process understanding. The challenge for this effort is to develop the requisite knowledge and process experience base, integrate these elements with an intelligent control system and sensor suite and transition this process into pilot-scale production in a microfactory within five years.

The design and manufacturing of next generation military systems will be performed in a complex, computer interconnected environment. Research will be conducted on cutting edge and exploratory materials with the results being actively compiled into topical knowledge bases for IPM production utilization at a later point in time. Development engineering would also include integration of process models with materials property databases and the topical knowledge bases from multiple researchers. Concurrent with this research function, manufacturability experiments will be conducted to validate new materials fabrication processes. Experiments such as the RSPD program and the development of the KAT system serve to demonstrate the viability of this real-time knowledge extraction approach.

The intelligent process controllers being created under IPM provide the necessary technology elements to enable high level functional design specification at workstations by applications engineers. Actual materials production will be tied to product design by this type of functional system compiler, since process knowledge and materials parameters will form an integral part of the systems's knowledge base. All manufacturing specifications will be sent directly to intelligent processors on the factory floor, which will coordinate component development and integration from raw materials purchase through final evaluation. The goal of moving this level of intelligent control and integration forward in the design and engineering specification process to reduce product development times. Thus, by creating and demonstrating the technology base for advanced materials development and production, the IPM program provides a critical linkage to the advanced manufacturing and production of advanced DoD systems.

Acknowledgements

Preparation of this manuscript was supported in part by the Defense Advanced Research Projects Agency (DARPA) and the Air Force Office of Scientific Research (AFOSR) under contract F49620–86–C–0036. The authors wish to express their appreciation to Dr. D. Ulrich of AFOSR, Dr. C. R. Green of the Army Research Office, Dr. H. N. G. Wadley of the National Bureau of Standards, Dr. R. T. Wood of General Electric, Dr. W. J. Pardee of Rockwell and Drs. T. A. Blaxton and C. B. Friedlander of The BDM Corporation.

References

1.   R. T. Wood, *Intelligent Processing Initiative for Gallium Arsenide Crystal Growth*, Second Quarterly Report, April 1987

2.   W. Loomis, et al, *Analytical Processing for Improved Composites (APIC), Final Report*, AFWAL–TR–81–4082 (1981)

3.   B. G. Kushner, R. A. Geesey, P. A. Parrish and S. G. Wax, "A Knowledge Acquisition Tool for the Intelligent Processing of Advanced Materials," *Proceedings of the 1986 TMS/ASME Conference on Applications of Artificial Intelligence to Materials Processing*, TMS/ASME Press, 1987.

4.   T. A. Blaxton and B. G. Kushner, "An Organizational Framework for Comparing Adaptive Artificial Intelligence Systems," *Proceedings of the 1986 IEEE–ACM Fall Joint Computer Conference*, IEEE Press, 1986.

INTELLIGENT PROCESSING OF POLYMERIC COMPOSITES*

B.R. Tittmann, W.J. Pardee and F. Cohen-Tenoudji

Rockwell International Science Center
Thousand Oaks, CA 91360

Abstract

The autoclave curing of graphite/epoxy is discussed in terms of an approach based on advanced sensors, computer modeling of the cure cycle, and AI/expert system control of the process variables. An ultrasonic viscosity sensor is used to determine the storage (real) and loss (imaginary) components of the shear modulus from which the high-frequency dynamic viscosity is calculated. This information forms an input into the modeling and simulation program to determine the state of the cure and to simulate the remainder of the cure cycle. The use of an AI/expert system controller allows the optimization of the cure cycle to minimize porosity and delaminations, to improve control of the resin's compaction, and to insure complete curing of the component.

* Work supported by Rockwell International Independent Research and Development Program.

Introduction

Polymer-based composite materials used in commercial and aerospace applications are often based on a matrix of thermocuring polymers which are usually cured in an autoclave. Under the effects of temperature, the polymer molecules grow into longer chains and crosslinks. The rate of the reaction is a complex function of the temperature and pressure, which depends on the thickness and geometry of the part being made, on the thermal equilibrium between the part and the mold, on the temperature of the atmosphere around the part, and/or the thermal mass of the autoclave. The monitoring of the viscoelastic properties of the resin provide important information on the state of the cure. Here, we present results for a high-temperature ultrasonic viscosity sensor. The resin's complex shear modulus at 1 MHz is derived from the measured reflection coefficient of shear wave pulses at the tool-resin interface. A special transducer-buffer assembly that operates at high temperature and provides a reference calibration signal has been developed. With this assembly, absolute determinations are made throughout the cure cycle of the storage (real) and loss (imaginary) components of the shear modulus from the latter of which the high-frequency dynamic viscosity is calculated. The information from the advanced sensors forms an important input into our computerized modeling and simulation program to determine the state of the cure and to simulate the remainder of the cure cycle. This program is an extension of the modeling programs developed by Lee, Loos and Springer (1) and allows the analytical simulation of the cure cycle, taking into account an arbitrary number of layers of the part and predicting when full compaction of these layers occurs. The use of a computer-aided system controller allows the optimization of the cure cycle to minimize porosity and delaminations, to improve control of the resin's compaction, and to ensure complete curing of the component. An approach is presented on how a system controller would take the information from the advanced sensors and the simulation program to arrive at the optimum cure cycle control parameters.

Advanced Sensors

The type of sensor technology needed for automated curing depends on the control philosophy. It is argued here that it is important to close the control loop on not just temperature, but also material properties. Most currently used autoclave systems give no feedback about the material state or degree of cure. The autoclave controller therefore cannot compensate for materials variability or the inevitable approximations in the models of the thermal, chemical and hydrodynamic response of the part-tool system. The minimum information needed to close the loop in resin curing is knowledge of the resin viscosity. This parameter and temperature together can be used to determine the glass transition temperature, a good measure of the degree of cure. The resin viscosity is also the critical material variable determining compaction.

Table I lists the principal techniques available for viscosity measurements, with a summary of their attractions and limitations. The dielectric techniques are limited by the ambiguity of the correlation between dielectric response and viscosity. Acoustic techniques have long been attractive for measurement of resin viscosity, porosity and compliance (2). A suitable acoustic technique can provide viscosity before gelation and compliance after gelation. The sharp rise in modulus (drop in compliance) following gelation is a direct measure of the degree of cure of the polymer network. A sensor that can detect degree of cure near the end of the cure cycle has obvious attractions in increasing autoclave throughput.

Most common acoustic measurements in composites have been longitudinal wave velocity measurements, although they are not simply related to the polymer viscosity. Since viscosity is intrinsically a relation between shear force and shear strain rate, it is best measured with shear waves. The shear modulus, or compliance, after gelation rises steeply as the network crosslinks, and so shear waves are sensitive to the degree-of-cure after gelation. Figure 1 illustrates schematically the acoustic interrogation of a curing polymeric composite. For a shear wave pulse impinging from the left, the tool (buffer)-resin interface has a complex reflection coefficient R given by

186

Table I. Viscosity Measurement Techniques

Technique	Parameters Measured	Deficiencies	Advantages	Production Applicability
Dielectric				
1. Audrey	• Capacitance (ϵ') • Dissipation factor (D)	• Noise reliability • Poor interpretability	• Has significance especially during processible portion of cure cycle	• Viewed with suspicion by many shops • Needs more development for reliability
2. G.D.	• Capacitance (ϵ') • Phase (θ)	• Dependent on external environment • Subject to setup conditions		
3. Ion Graphing	• DC resistivity			
4. Microdielectrometry	• Temperature • Gain and phase of feedback circuit which is related to ϵ' and D	• Invasive	• Small size • Inexpensive • Reliable • Rugged	• Microchip technology
Acoustic				
1. Shear Wave Transmission	• Attenuation α_s • Velocity ν_s	• High loss of signal	• Acoustics is non-invasive • Directly related to viscosity	
2. Shear Wave Pulse Reflectivity	• Complex reflection coefficient	• Required well-defined acoustic interface		• High-temperature ultrasonic transducers technology
3. Shear Wave CW Resonance	• Quality factor and velocity			
4. Longitudinal Wave Transmission	• Attenuation α_L • Velocity ν_L	• Not easily related to viscosity	• Gives microporosity, gelation	
5. Acoustic Waveguide	• Signal amplitude	• Invasive	• Detects degree of cure • Not directly related to viscosity	• Needs more development

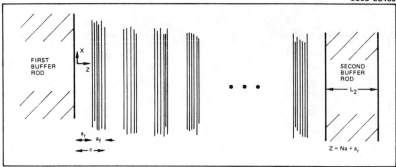

Figure 1 - Acoustic interrogation of a fiber-reinforced composite. The laminae are of approximately equal separation a = a (resin) + a (fibers). A shear wave is incident from the left; the second (right) buffer rod is optional.

$$R = \frac{Z_R - Z_B}{Z_R + Z_B}; \quad Z_B = \sqrt{\rho_B G_B};$$

$$Z_R = \sqrt{\rho_R G_R} = \sqrt{\rho_R G_R' + i\rho_R G_R''}$$

where Z_R, Z_B are the shear wave acoustic impedances, ρ_R and ρ_B the densities, G_R and G_B the shear moduli of the resin and buffer, respectively.

The dynamic viscosity (before gelation) is $\eta \cong G''/\omega$, and G_R' and G_R'' can be obtained from these three equations in terms of measured or known values of R, ρ_B, ρ_R (approximately constant) and G_B. We have recently demonstrated the usefulness of this approach by development of a high-temperature sensor which was successfully used to monitor the effective dynamic viscosity of Hercules 3051-5A during the cure cycle in a research autoclave. Figure 2 shows the magnitude of the reflection coefficient as a function of time into the cure. Figures 3a and 3b show a comparison between the dynamic viscosity of 1 MHz and the viscosity obtained at 10 Hz with a conventional

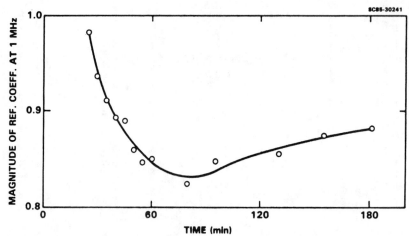

Figure 2 -Magnitude of ultrasonic reflection coefficient as a function of time into the cure cycle.

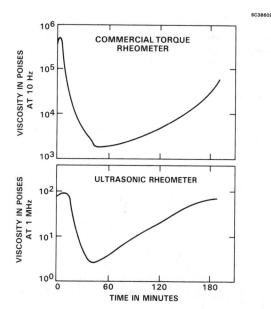

Figure 3 - Dynamic "effective" viscosity (a) and quasi-state "effective" viscosity (b) measured by ultrasonics (1.0 MHz) and by torque-rheometric techniques, respectively, of Hercules 3501-5A during simultaneous measurements as a function of time into the cure cycle.

torque viscometer. Both measurements were made simultaneously by modifying a torque viscometer sample holder to allow simultaneous measurements of the shear wave reflection coefficient while carrying out the conventional torque measurements. Both curves show the same strong decrease and subsequent minimum in viscosity at the same point in the cure cycle. Similar rises in viscosity are seen as gelation occurs. Note that the viscosity changes are several orders of magnitude. Because the mechanism is a relaxation process, the absolute values and net changes cannot be expected to be the same between the two measurements, which are a factor 10^5 apart in frequency. Aside from these considerations, the performance of this type of sensor appears promising for monitoring the local viscosity of a fiber-reinforced composite during an autoclave cure cycle. The employment of the sensor is visualized as being built into the curing tool to allow contact with the resin surface of a component at a few critical points to optimize curing, particularly where there are changes in thickness or unusual co-cured structures. Note also that the measurements of the magnitude of R values for the shear modulus provide valuable information about when the cure is complete.

Modeling of Cure Cycle

The fundamental controllable variable in polymer curing is heat. Uniform temperature throughout the component regardless of thickness or placement within the autoclave requires that the laminate temperature be kept relatively uniform, that it not exceed a maximum determined by the specific possible reaction kinetics, and that the temperature be such that the reactions occurring lead to acceptable final values of properties such as glass transition temperature. Loos and Springer (3) write the model energy balance in the form

$$\partial(\rho c T)/\partial t = \partial/\partial Z \, [k \partial T/\partial Z] + \rho h \quad \dot{\alpha}$$

189

where ρ and c are the density and specific heat, respectively, of the laminate, T is the temperature, k is its thermal conductivity normal to the layup plane, h is the heat of reaction, $\dot{\alpha}$ is the reaction rate, t is time, and Z is distance from the tool in the direction normal to the layup plane. Loos and Springer define the degree of reaction α as the heat evolved at time t divided by the total heat of reaction $\alpha(t) = h(t)/h$. The time derivative of $\dot{\alpha}$ describes the rate of heat evolution with time.

For the solution of the differential equation above, Lee, Loos and Springer (1) fit differential scanning calorimetry results to an eight-parameter expression of the form $\dot{\alpha} = f(T,\alpha)$. Curing of the standard 350°F curing epoxy resin (TGMDA with curative DDS and catalyst BF_3) involves several alternative reaction paths with quite different reaction enthalpies. Therefore, changing the temperature changes the relative rates of these reactions, and thus the resulting networks and viscosity. This is the mechanism by which cure path influences final mechanical properties. The next step in model sophistication is to write the rates f_1 and f_2 of these two reactions in terms of the fraction x_1 of epoxy consumed and the fraction x_2 of curative consumed. Then, $h\dot{\alpha} = h_1 f_1 + h_2 f_2$, where h_1 and h_2 are the corresponding heats of reaction, respectively. These equations can now be manipulated to predict (4) the glass transition temperature. They also allow the prediction of the viscosity through the Williams, Landel, Ferry (WLF) relation

$$\ln \eta(T) = \ln \eta(T_g) + 40.7 (T-T_g)/[51.6 + |T-T_g|]$$

Springer (5) has shown that compaction occurs one ply at a time and that there are two regions separated by a sharp boundary, one fully compacted and the other having the original resin content.

Using Darcy's law to describe resin flow velocity, Loos and Springer derive for the rate dh_b/dt at which resin rises in the bleeder as a result of compaction

$$\phi_b \frac{dh_b}{dt} = v$$

where ϕ_b is the fraction of bleeder volume available to the resin. Between ply compactions, this equation has the form

$$(A + Bh_b) \frac{dh_b}{dt} = P_o - P_b$$

where P_o and P_b are the pressures in the uncompacted and compacted regions, respectively. A changes each time another ply compacts. This model, called the Loos-Springer cure simulation model, allows computer prediction of composite curing. A good example is provided by the 10^4 ply graphite/epoxy B-1B vertical stabilizer skin, whose cure was simulated by one of us (W. Pardee) using this code. Figure 4 shows cure according to a traditional minimum specification and the resulting failure to reach full compaction. Using the longest normal autoclave cycle (Figure 5), the skin was still only 57% compacted. The simulation finally predicted full compaction for the cycle shown in Figure 6.

Real-Time Control System

Here, we describe an approach for a real-time control system. An overview of the functions of the system is shown in Figure 7. At the lowest level, it includes the autoclave itself; at the highest level, the system host interacts with both the user and the factory supervisor, as well as generating a specific temperature and pressure sequence based on measured values of temperature T, pressure P, viscosity η and porosity δ. The major objectives of this design are to:

190

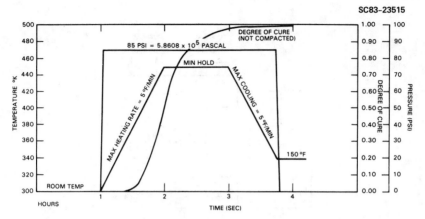

Figure 4 - Minimum cure cycle.

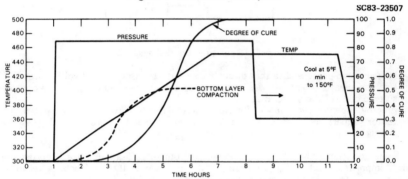

Figure 5 - Maximum cure cycle.

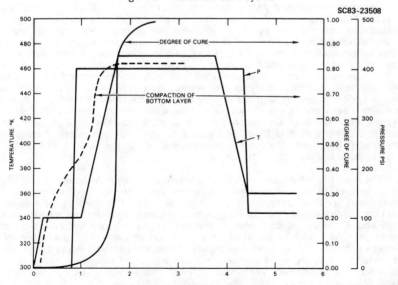

Figure 6 - Optimized cure cycle.

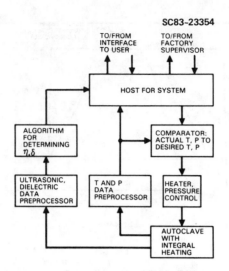

Figure 7 - Real-time control system.

1. Measure and process data that directly correlate to viscosity and void content.
2. Use the measurements of viscosity and porosity, as well as temperature and pressure to control the autoclave.
3. Employ the cure cycle model for each particular autoclave load constructed in the factory supervisor to generate the desired sequence of temperature and pressure.
4. Support user interface and data base updating actions.

The functions labeled "Heater, Pressure Control", "T and P Data Preprocessor", and "Comparator: Actual F,P to Desired T,P" are essentially the feedback servo loop for control of temperature T and pressure P.

The functions labeled "Algorithms for Determining η and δ" and "Ultrasonic, Dielectric Data Preprocessor" include ultrasonic and dielectric sensor data acquisition, preprocessing, and the necessary calculations to provide estimates of the viscosity and porosity at each sensor position. For example, the ultrasonic sensor data acquisition system includes, besides the transducers and their associated pulser/receivers, a multiplexer and an array processor, as shown in Figure 8. At run time, the multiplexer receives commands from the array processor and selects a transducer for data acquisition. The array processor analyzes the received data after conditioning and digitization. A typical waveform digitizer is designed to handle 18 megasamples/s. Assuming the bus rate between the array processor and the waveform digitizer is 1 Mbyte/s (burst rate) and there are 1024 data points per waveform, it will take approximately 1 ms to transmit the digitized data from the digitizer to the array processor. Adding in the transducer pulsing time, the control time and the handshake time, it will take about 1.5 ms to pulse and process one transducer.

The most important function of the topmost level of the real-time controller, the host, is to execute the control algorithm for generating the values of temperature and pressure that are desired for the next time segment, based on the cure cycle model for each load and the actual values of temperature, pressure, viscosity and porosity reached at the end of the last time segment. The time segment length will match the cycle time of the CAPS 210. Second, the host will respond to user requests and allow on-line changes to the cure cycle model if necessary. Third, the host will report T, P, viscosity and porosity values for each time segment to the factory supervisor for entry in the data base. Finally, the host will handle all the startup and close-out control communications required by the user, the autoclave and the supervisor.

192

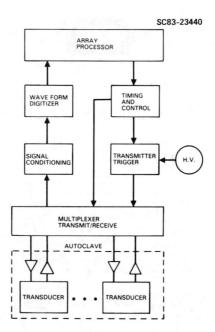

SC83-23440

Figure 8 - Ultrasonic sensor data acquisition subsystem.

AI/Expert Systems Control

The use of an AI/Expert system controller allows the optimization of the cure cycle. Intelligent modification of the materials process is achieved by the cooperation of three materials and process rule system modules whose functions are: (1) intelligent interpretation of sensor data to make the best possible estimate of the current state of the material, including inferences about unmeasured quantities; (2) evaluation of cure plans, either the current one or a proposed alternative one in light of those state estimates, to assess whether quality of the cured parts will be acceptable and whether the plan is efficient; and (3) planning of new cure cycles to solve problems anticipated by the expert.

These three agents, Estimator, Evaluator and Planner, must be present in any intelligent materials process control system in one form or another, although their relative sophistication may vary somewhat from process to process. One of the attractions of the composite cure problem as a prototype is that it appears that none of the three is either trivial or unreasonably difficult. The functions of these three agents correspond very roughly to diagnosis, prognosis and prescription.

To coordinate their functions and to manage their backlog queues to best allocate system resources, their communication is through a fourth agent, System Manager, which also manages communication with the real-time simulation module needed by both Estimator and Evaluator, with the autoclave and the operator. System Manager will also transmit explanations of system reasoning to the operator. To prioritize the tasks of the three materials and process agents, System Manager must also have some domain knowledge.

This architecture is shown schematically in Figure 9. The AI software could be implemented in Intellicorp's Knowledge Engineering Environment (KEE) running on LMIs Lambda 2 x 2/plus Lisp machine with additional numeric processor. The anticipated separation of functions between the numerical and symbolic environments is illustrated

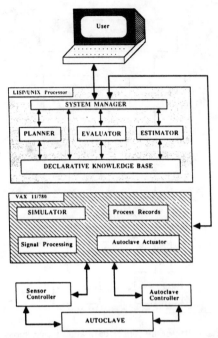

Figure 9 - Schematic of hybrid symbolic/numeric autoclave control.

in Figure 9. The LMI machine is unique in providing high-speed symbolic computation and a bus architecture that permits very high-speed communication with a numeric processor. In addition, the machine can communicate at Ethernet speed with the VAX 11/780 or 11/750 where most of the low-level data acquisition and processing as well as the large numerical simulation will be performed. In common with all modern Lisp machines, the LMI has a high resolution, bit-mapped display, and an optical mouse to make possible a high-speed, broad-band user interface. This architecture is very well suited to development because it provides power and flexibility in both symbolic and numeric domains. It is composed of generally available commercial components and software, so results can be widely adapted.

No materials process can be controlled intelligently without real-time knowledge of material state. For the composite cure process, three fundamental kinds of material state sensors are possible: microdielectric chips for temperature and dielectric response (related to viscosity); ultrasonic transducers for viscosity and elastic modulus; and ultrasonic transducers for porosity. These three techniques have been shown at least plausible, but are not yet used in production. Together, they provide a fairly comprehensive description of material state, require nontrivial but manageable signal processing, and are made more useful by heuristic reasoning about related properties not directly measured. These measurement techniques, combined with their high-level interpretation by Estimator, comprise a sophisticated stand-alone expert monitoring module of very general significance. The temperature and pressure of the autoclave and the pressure on the individual bags can be monitored as in conventional autoclave cure.

References

1. W.I. Lee, A.C. Loos and G.S. Springer, "Curing of Epoxy Matrix Composites," Comp. Matls., 16 (1982), 510.

2. G.A. Sofer and E.A. Hauser, J. Polym. Sci. 8 (1952), 6.

194

3. A.C. Loos and G.S. Springer, <u>J. Comp. Matls.</u> 17 (1983), 135.

4. D.H. Kaelble, <u>CAD/CAM Handbook for Polymer Composite Reliability</u> (Final Report for the period Nov. 1, 1980 to Nov. 1, 1982, p. 91, U.S. Army Research Office, Contract NO. DAAG-29-80-C-0137, March 1983).

5. G.S. Springer, <u>J. Comp. Matls.</u> 16 (1982), 400.

MODELLING AND MEASURING THE CURING AND CONSOLIDATION

PROCESSES OF FIBER-REINFORCED COMPOSITES

Alfred C. Loos
Department of Engineering Science and Mechanics
Virginia Polytechnic Institute and State University
Blacksburg, VA 24061

David E. Kranbuehl
Department of Chemistry
College of William and Mary
Williamsburg, VA 23185

W. T. Freeman, Jr.
Applied Materials Branch
NASA-Langley Research Center
Hampton, VA 23665

Abstract

A processing model and an in-situ sensing measurement technique are
discussed for use in automated composite processing and quality control.
Frequency-dependent electromagnetic sensors (FDEMS) were used to
simultaneously measure the resin viscosity at four locations inside a thick
section 192-ply graphite-epoxy composite during autoclave cure. The
viscosity measurements obtained using the sensors were compared with the
viscosities calculated using the cure process model. Good overall agreement
was observed between the predicted and sensor measured viscosity at each of
the four ply locations.

Introduction

Composite laminates constructed from continuous fiber-reinforced thermosetting matrix prepreg materials are fabricated by laminating multiple plies into the desired shape and then curing the material in an autoclave by simultaneous application of heat and pressure. Elevated temperature applied during cure initiates and maintains matrix chemical reactions which cause the desired molecular structure changes while applied pressure compacts and consolidates the prepreg plies. The magnitude and duration of process temperatures and pressures coupled with the state and cure properties of the matrix critically influence the final physical and mechanical properties of the composite. For many high performance resin systems empirically derived process cycles often lead to part rejection rates of 10 to 70%.

The large number of material properties and processing parameters that must be specified and controlled during cure of a composite laminate make trial-and-error procedures to determine the cure cycle extremely inefficient. Effective analytical models are clearly a far superior alternative for determination of optimum cure cycles. In-situ sensors which can measure critical processing properties, such as viscosity and reaction advancement, are essential for verifying the predictions of the analytical model, refining the model, and testing under conditions which cannot be fully represented analytically. Ultimately both the analytical model and the sensing technology are required for a closed-loop intelligent production control system.

In this paper a model of the curing and consolidation processes of fiber-reinforced composites is presented. The model can be used to determine the appropriate cure cycle for a given application which results in a composite that is uniformly cured to the desired resin content in the shortest amount of time.

The results of work on the development of frequency dependent electromagnetic techniques as a "smart" sensor for quantitative nondestructive material evaluation and intelligent, closed-loop process control systems are reported. Use of frequency dependent electromagnetic sensors (FDEMS) for monitoring the curing process of fiber-reinforced composites will be discussed.

Experimental

Composite specimens for this program were fabricated from Hercules AS/3501-6 graphite-epoxy unidirectional prepreg tape. The resin content of the prepreg was 35% by weight and the fiber areal weight was 149 g/m^2.

Lay-up preforms 32-plies thick and 30.48 cm by 30.48 cm were fabricated using a cross-ply stacking sequence. Each 32-ply lay-up was precompacted at room temperature under a vacuum of 635-mm (25-in.) mercury for 20 minutes and then were stored in sealed polyethylene bags at -18°C. A 192-ply composite panel was fabricated by combining six of the precompacted 32-ply modules as shown in Fig. 1. No. 30 gage iron constantan thermocouples and Dek Dyne FDEMS electromagnetic sensors were placed in each of the four locations as close as possible to the geometric center of the lay-up as indicated in Fig. 1.

The 192-ply composite lay-up was bagged as shown in Fig. 2 and placed in a 4 foot diameter by 8 foot autoclave for cure according to the manufacturer's recommended cure cycle (Fig. 3). During cure, temperature and complex permittivity inside the composite and autoclave air temperature

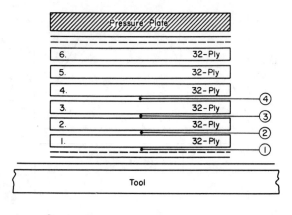

Location of thermocouples and dielectric sensors

① Tool surface
② 32 nd Ply
③ 64 th Ply
④ 96 th Ply

Figure 1 - Schematic of the 192-ply graphite-epoxy composite showing the locations of the thermocouples and FDEMS sensors.

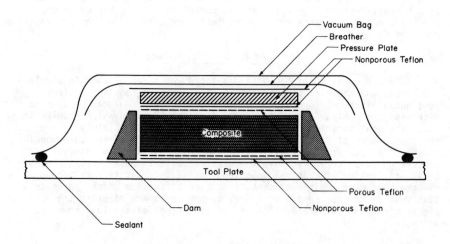

Figure 2- Bagging schematic.

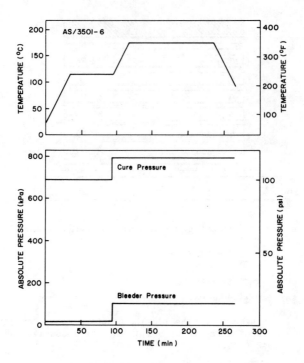

Figure 3 - Manufacturer's recommended cure cycle for
Hercules AS/3501-6 prepreg.

were recorded by a Hewlett-Packard 3497A data acquisition system controlled
by a Hewlett-Packard 9836 computer. Descriptions of the frequency dependent
electromagnetic impedance measurement instrumentation have been previously
reported (1-5).

Cure and Consolidation Model

A cure process model has been developed which can be used to simulate
the curing and consolidation processes of composites constructed from
continuous fiber-reinforced prepreg materials. The model can be used to
determine the optimum cure cycle for a given application which results in a
composite that is void free, and cured uniformly and completely in the
shortest amount of time. Other important applications of the cure model
include: a) performing sensitivity studies to identify the material and
processing parameters that have the greatest effect on the curing process;
b) identification of critical locations inside a composite part for
placement of cure monitoring sensors; c) identification of the parameters
that must be controlled and monitored during cure; and d) using the model as
a cure simulator for the development of real-time, closed-loop computer
software.

A detailed description of the model has been previously reported in
references 6-8. Only a brief summary of the information that can be
generated by the model will be presented here.

200

The model relates the cure cycle and prepreg properties to the thermal, chemical, and rheological processes occurring in the composite during cure. For a flat plate composite, the model can be used to calculate the following parameters for a specified process cycle:

a) The temperature distribution inside the composite;
b) the temperature drop across the tooling, bleeders, and bagging material;
c) the degree of cure of the resin as a function of position and time;
d) the resin viscosity as a function of position and time;
e) the resin flow from the composite and the fiber/resin distribution as a function of time;
f) the number of prepreg plies that are fully compacted as a function of time;
g) the void sizes, temperatures, and pressures inside the voids as functions of void location and time;
h) the residual stress distribution inside the composite after cure; and
i) the thickness and mass of the composite as a function of time.

Different aspects of the model have been verified experimentally in previous investigations (9,10). In particular, experiments were performed measuring the temperature distribution through the thickness and the resin flow out of composites during cure. Composite test specimens were fabricated from graphite-epoxy prepreg with different ply stacking sequences, ply thicknesses, and length-to-width dimensions. Results of these studies indicate that the model describes adequately the temperature distribution and resin flow of flat plate graphite-epoxy laminates.

The development of in-situ cure-monitoring sensors will make it possible to measure the viscosity of the resin inside a composite during cure. These measurements can be compared with the resin viscosity calculated using the cure-process model to assess the validity of the model. Sensor measurements of the degree of cure and comparison with the model's predictions will be addressed in a subsequent report.

Electromagnetic Sensing Theory

Measurements of the complex impedance from Hz to MHz were used to calculate the complex permittivity

$$\epsilon^* = \epsilon' - i\epsilon'' \tag{1}$$

at particular positions in the 192 ply lay-up. This calculation is possible when using a probe whose geometry is invariant over all measurement conditions. Both the real and the imaginary parts of ϵ^* can have an ionic and dipolar component (11).

$$\epsilon' = \epsilon'_d + \epsilon'_i$$
$$\epsilon'' = \epsilon''_d + \epsilon''_i \tag{2}$$

The dipolar component arises from diffusion of bound charge or molecular dipole moments. The frequency dependence of the dipolar component may be represented by the Cole-Davidson function:

$$\epsilon_d^* = \epsilon_\infty + \frac{\epsilon_0 - \epsilon_\infty}{(1+i\omega\tau)^\beta} \tag{3}$$

where ϵ_0 and ϵ_∞ are the limiting low and high frequency values of ϵ, τ is a characteristic relaxation time, and β is a parameter which measures the distribution in relaxation times. The dipolar term is generally the major component of the dielectric signal at high frequencies and in highly viscous media.

The ionic component, ε_i^*, often dominates ε^* at low frequencies, low viscosities and/or higher temperatures. The presence of mobile ions gives rise to localized layers of charge near the electrodes. Since these space charge layers are separated by very small molecular distances on the order of A^0, the corresponding space charge capacitance can become extremely large, with ε' on the order of 10^6. Johnson and Cole, while studying formic acid, derived empirical equations for the ionic contribution to ε^* (12). In their equations, these space charge ionic effects have the form

$$\varepsilon_i' = C_0 Z_0 \sin \frac{(n\pi)}{2} \omega^{-(n+1)} \left(\frac{\sigma}{8.85 \times 10^{-14}}\right)^2 \qquad (4)$$

where $Z^* = Z_0(i\omega)^{-n}$ is the electrode impedance induced by the ions and n is between 0 and 1 (11,12). The imaginary part of the ionic component has the form

$$\varepsilon_i'' = \frac{\sigma}{8.85 \times 10^{-14} \omega} - C_0 Z_0 \cos \frac{(n\pi)}{2} \omega^{-(n+1)} \left(\frac{\sigma}{8.85 \times 10^{-14}}\right)^2 \qquad (5)$$

where σ is the conductivity (ohm^{-1} cm^{-1}), an intensive variable, in contrast to conductance $G(\text{ohm}^{-1})$ which is dependent upon cell and sample size. The first term in Eq. (5) is due to the conductance of ions translating through the medium. The second term is due to electrode polarization effects. The second term, electrode polarization, makes dielectric measurements increasingly difficult to interpret and use as the frequency of measurement becomes lower.

Analysis of the frequency dependence of $\varepsilon^*(\omega)$ in the Hz to MHz range is, in general, optimum for determining both σ and τ. In turn, the ionic parameter, σ and the dipolar parameter, τ are directly related on a molecular level to the rate of ionic translational diffusion and dipolar rotational mobility (11).

Results and Discussion

The frequency dependence of the real and the imaginary parts of the complex permittivity are used to determine σ and τ, the ionic and dipolar contributions to ε^*. These parameters are measured during cure as a function of time. They reflect the mobility of the ions and dipolar groups, that is their molecular rates of translational and rotational diffusion. Properly used, these parameters which measure molecular mobility can be used to quantitatively measure the viscosity.

The correlation during cure of $-\log(\sigma)$ and $\log(\eta)$ (viscosity) and the ability to use frequency dependent $\varepsilon^*(\omega)$ measurements to determine σ, thereby accurately detecting points of maximum flow during cure, was first shown several years ago (2,3). The qualitative correlation of simultaneous measurements of $-\log(\sigma)$ and $\log(\eta)$ is demonstrated in Fig. 4. The quantitative relationship of $\log(\sigma)$ and $\log(\eta)$ is shown for an isothermal cure in Fig. 5. To a first approximation one might expect a plot of

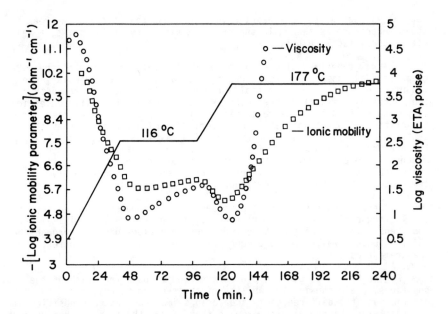

Figure 4 - Simultaneous measurement of specific conductivity and viscosity during cure for catalyzed epoxy.

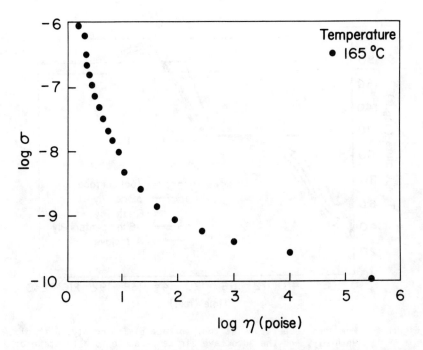

Figure 5 - Log σ vs. log η for uncatalyzed epoxy during isothermal cure.

log (σ) vs log (η) to be linear. Figure 4 shows that there is a break in the η dependence as the resin approaches a viscosity of 10^3 poise, a value often associated with gel.

The raw experimental data obtained during autoclave cure of the 192-ply graphite-epoxy composite are shown in Figs. 6-9. Figure 6 displays the measured temperatures on the tool surface ply, the 32nd ply, the 64th ply, the center ply as well as the autoclave air temperature. Differences in temperature of 10 to 20 degrees exist over extended lengths of time during the critical flow-consolidation cure periods. Figure 6 shows that the center (96th ply) ply temperature significantly lags the tool surface ply temperature until 10 minutes into the final hold. At this point the exothermic reaction drives the temperature rapidly above the autoclave air temperature and the center ply temperature exceeds the surface ply temperature throughout the remainder of the cure process.

Figures 7 and 8 show the raw values of ε'' scaled by the frequency for the 11 frequencies measured during cure. The noise level is minimal with good sensitivity showing that neither the connections passing through the thick autoclave wall nor the electrical surges during heating affect the measurement capability.

Figure 9 is a comparison plot of the raw ε' data from each of the 4 ply positions, again showing the ability of the sensing technique to discriminate the differences in the resin's impedance at each position with a sensitivity which equals or exceeds that of the thermocouple measurements. Most importantly ε' and ε'' reflect the actual cure properties of the resin.

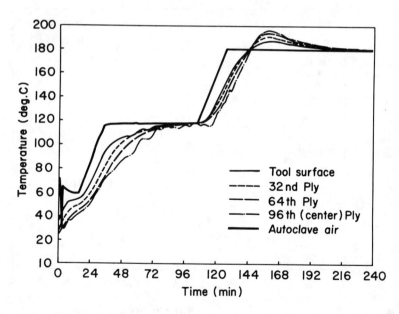

Figure 6 - Measured temperature at tool surface ply, 32nd ply, 64th ply, 96th ply, and the autoclave air temperature as a function of time.

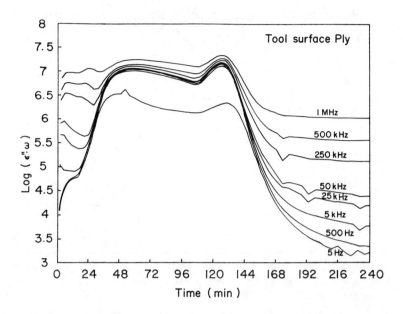

Figure 7 - Log (ϵ'' . ω) versus time. Measurements obtained from the FDEMS sensor located at the tool surface ply.

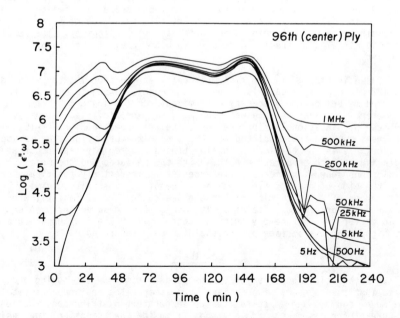

Figure 8 - Log (ϵ'' . ω) versus time. Measurements obtained from the FDEMS sensor located at the 96th (center) ply.

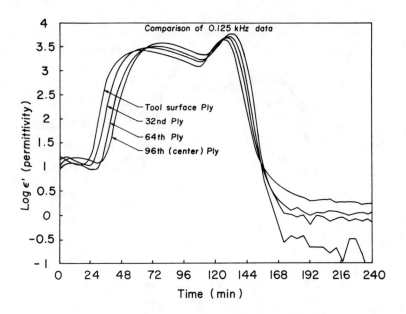

Figure 9 - Comparison plot of the log ε' versus time measured at the four ply positions inside the composite.

A series of rheometer experiments measured simultaneously the viscosity and the value of $\varepsilon^*(\omega)$ for temperature ramp-hold sequences similar to those observed in the autoclave cure. Using Eq. (5), the value of σ, reflecting ionic diffusion was determined and correlated with viscosity. The results of these correlation studies were used to quantitatively relate σ to viscosity.

Shown in Fig. 10 is the resin viscosity at the four locations inside the 192-ply composite calculated using the model. Shown in Fig. 11 are the measured values of the viscosity determined from the ionic mobility parameter σ. Comparisons between the measured and predicted viscosity at the four ply locations inside the composite are shown in Figs. 12-15. The agreement between the predicted and the sensor measured viscosity is good particularly in the critical high flow processing regions that occur at the beginning of each temperature hold. Both the measured and the model predictions show significant differences for different ply positions for both the time of maximum flow and the value of the viscosity minima. Time differences of up to 20 minutes between the center and surface ply are measured and predicted. The viscosity minima are measured and predicted to be lower for the surface ply than the center ply in the low temperature hold but higher for the surface ply in the high temperature hold.

Conclusions

A mathematical model which can be used to simulate the curing and consolidation processes of autoclave cured composites has been presented. The model can be used to calculate the temperature distribution, the degree of cure of the resin, the resin viscosity inside the composite, the resin flow, the potential for void formation, and the residual curing stresses for flat plate composites cured by a specified cure cycle.

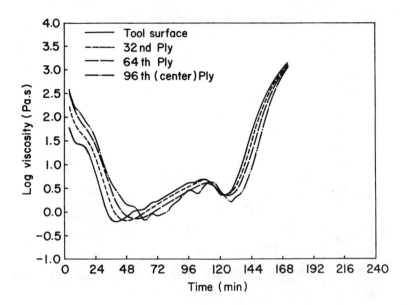

Figure 10 - Resin viscosity as a function of time at the four ply
locations. Results calculated using the cure model.

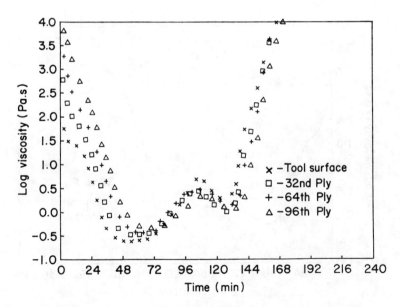

Figure 11 - Measured resin viscosity as a function of time at the
four ply locations. Measurements obtained using the
FDEMS sensors.

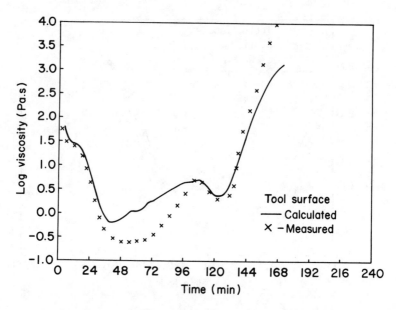

Figure 12 - Comparison between the resin viscosity measured using the FDEMS sensor and calculated by the model at the tool surface ply.

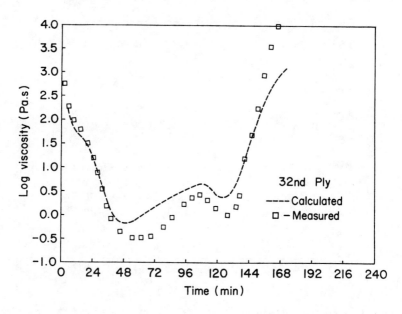

Figure 13 - Comparison between the resin viscosity measured using the FDEMS sensor and calculated by the model at the 32nd ply.

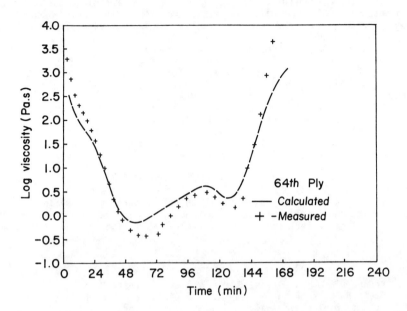

Figure 14 - Comparison between the resin viscosity measured using the FDEMS sensor and calculated by the model at the 64th ply.

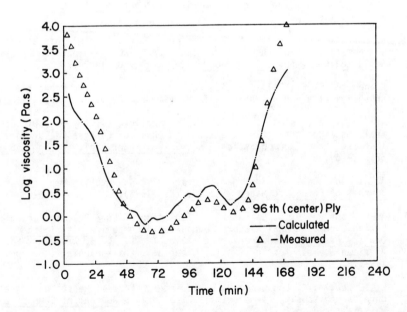

Figure 15 - Comparison between the resin viscosity measured using the FDEMS sensor and calculated by the model at the 96th (center) ply.

A frequency dependent electromagnetic sensing technique for observing and measuring molecular diffusion rates and thereby the resin viscosity of a composite cured in a production autoclave has been developed.

A number of first time advances in the use of both the model and the sensing techniques have been reported, including:
- Use of a sensing technique to make simultaneous viscosity and ionic mobility measurements and thereby use the frequency dependence of the complex impedance to measure viscosity.
- Use of a sensing technique to measure the resin viscosity in a thick 192-ply graphite-epoxy laminate as a function of position while being cured in an autoclave.
- Use and verification of a model to predict the viscosity in a thick laminate at various ply positions.

Acknowledgements

This work was made possible through the support of the National Aeronautics and Space Administration-Langley Research Center grant no. NAG1-343 with Virginia Tech and grant no. NAG1-237 with the College of William and Mary.

References

1. Kranbuehl, D. E., Delos, S. E., Yi, E. C., Hoff, M. S. and Whitham, M. E., "Analysis of the Polymerization of Amine-Based Epoxy Resins by Dynamic Dielectric Analysis," ACS Polym. Mater. Sci. and Eng., 53, 191 (1985).

2. Kranbuehl, D. E., Delos, S. E., Jue, P. K., Jarvie, T. P., and Williams, S. A., "Dynamic Dielectric Characterization of Thermosets and Thermoplastics Using Intrinsic Variables," National SAMPE Symp. Ser., 29, 1251 (1984).

3. Kranbuehl, D. E., Delos, S. E., and Jue, P. K., "Dynamic Dielectric Characterization of the Cure Process: LARC-160," National SAMPE Symp. Ser., 28, 608 (1983); SAMPE Journal, 19 (4), 18 (1983).

4. Kranbuehl, D. E., Delos, S. E., Jue, P. K., "Dielectric Properties of the Polymerization of an Aromatic Polyimide," Polymer, 27, 11 (1986).

5. Kranbuehl, D., Delos, S., Hoff, M., and Weller, L., "Dynamic Dielectric Analysis: Monitoring the Chemical and Physical Changes During Cure of Thermosets and Thermoplastics," ACS Polym. Mater. Sci. and Eng., 54, 535 (1986).

6. A. C. Loos and G. S. Springer, "Curing of Epoxy Matrix Composites," Journal of Composite Materials, 17, 135 (1983).

7. A. C. Loos and G. S. Springer, "Calculation of Cure Process Variables During Cure of Graphite-Epoxy Composites," Composite Materials: Quality Assurance and Processing, ASTM STP 797, C. E. Browning. Ed., American Society for Testing and Materials, Philadelphia, 1983, pp. 119-118.

8. W. I. Lee, A. C. Loos, and G. S. Springer, "Heat of Reaction, Degree of Cure, and Viscosity of Hercules 3501-6 Resin," Journal of Composite Materials, 16, 510 (1982).

9. A. C. Loos and W. T. Freeman, Jr., "Resin Flow During Autoclave Cure of Graphite-Epoxy Composites," High Modulus Fiber Composites in Ground Transportation and High Volume Applications, ASTM STP 873, D. Wilson, Ed., American Society for Testing and Materials, Philadelphia, 1985, pp. 119-130.

10. A. C. Loos, "Modeling the Curing Process of Thermosetting Resin Matrix Composites," Review of Progress in Quantitative Nondestructive Evaluation, Vol. 5B, ed. D. O. Thompson and D. E. Chimenti (New York, N.Y.: Plenum Press, 1986), 1001-1012.

11. Hill, N., Vaughan, W., Price, A., and Davis, M., Dielectric Properties and Molecular Behavior, Van Nostrand, London (1969).

12. Johnson, J. and Cole, R., "Dielectric Polarization and Relaxation," J. Am. Chem. Soc., 73, 4536 (1985).

CHARACTERIZATION AND CONTROL OF CONTINUOUS CASTING PROCESSES

R. C. Sussman and J. R. Cook

Armco Research, Middletown Ohio 45043

Abstract

The continuous casting process has become the key to enhanced quality and improved competitiveness for the modern steel industry. This process has the ability to continuously produce material in sizes and shapes appropriate to a wide range of subsequent processing steps. Recent innovations in continuous casting machine design allow the process to be more closely coupled to successive processing steps and have resulted im markedly improved processing efficiencies. This capability has been enhanced by advances in metallurgical and processing knowledge, and in the ability to sense and control the processing parameters of interest. In many cases those parameters cannot be directly measured and must be inferred by indirect means. This paper will discuss the problems and requirements for control in both conventional and thin strip casting processes. Recent effort on the collaborative work (American Iron & Steel Institute, National Bureau of Standards, Department of Energy) to develop a device to monitor the progress of the solidification process and to control the continuous caster will also be outlined.

213

Introduction

A hundred years elapsed after Sir Henry Bessemer conceived and pa-
tented the concept of continuous casting before the first machines were
installed in melt shops of steelworks; nevertheless, continuous casting is
now the choice for new steelmaking plants. The conversion of older facilities is taking place as rapidly as the generation of capital will permit.
The quantity of steel produced as continuous cast product has grown rapidly
in the last 20 years, as shown in Figure 1. In the U.S., where tight capital has limited modernization, continuous casting already exceeds 50% of
steel production; in Japan, that proportion is greater than 90% (1,2).

The economic advantages of continuous casting are well known. The
alternative process is ingot casting, in which molten steel is poured into
molds for partial or complete solidification. The molds are removed and
the ingots placed in soaking pits to equilibrate the steel temperature before rolling to shape in a primary mill. The ingot process is lengthy and
requires careful control of the casting, reheating, soaking, and rolling
processes. The continuous casting process eliminates or combines those
steps into a single process resulting in an as-cast shape suitable for secondary rolling (Figure 2).

In conventional continuous casting, metal is poured from the transfer
ladle into a tundish and then into a water-cooled copper mold in which a
shell of solidifying material serves to contain the liquid steel. A great
deal of attention is paid to the mold flux which provides both the lubricity and assists heat transfer. The strand is contained by closely spaced
rolls while continuing to cool and solidify. This secondary cooling segment removes energy both by water spraying and direct roll contact. The
steel solidifies from the outer periphery and the center of the product may
remain unsolidified well beyond the last containment sections of the
machine. Cooling along the various segments of the casting machine must be
precisely controlled to avoid sensitive temperature ranges at points of
high stress. The can be accomplished by controlling the strand velocities
and modulating spray cooling rates.

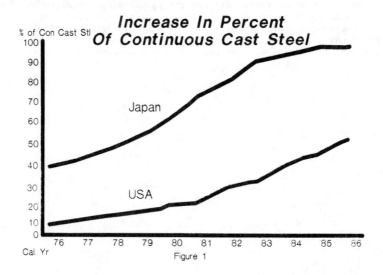

Figure 1

214

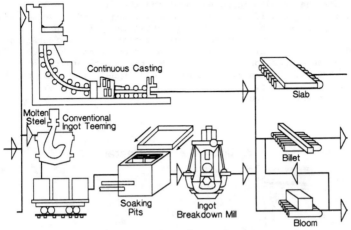

Continuous Casting

Slab

Molten Steel

Conventional Ingot Teeming

Billet

Soaking Pits

Ingot Breakdown Mill

Bloom

Figure 2

The benefits of continuous casting over ingot casting can be considerable (3):

1. Continuous casting product quality is improving steadily as methods are developed to minimize indigenous formation, float out existing inclusions in the tundish and mold and reduce the propensity for cracking. Modern continuous casters can now produce superior quality than equivalent grades produced by ingot casting.

2. Yield increases of 7-20% and energy savings of up to 1 million BTU's per ton cast are achievable over ingot processing through the elimination of ingot processing steps. The soaking pits, in particular, can be a very fuel-consumptive operation, although alternative liquid center charging and rolling techniques have evolved to extend the economic life of some modern integrated ingot processing facilities.

3. Improved labor productivity and marked reductions in in-process inventory have been achieved by the combining of processes and the simplification and rationalization of product lines.

4. The elimination of the need for ingot pouring aisles, stripping equipment, soaking pits and primary mills also results in reduced maintenance costs and eliminates the risk of air pollution during teeming.

Types of Casters

Although continuous casting refers to processes in which the melt is solidified in a continuous strand, considerable differences exist among the various machines. These differences are primarily geometric and are dictated by the requirements of the product, capital constraints, and marketing decisions. The machines can be categorized into common technological types:

1. Vertical slab casters commonly cast shapes 125-305 mm (5-12 in.)
    thick and at least 810 mm (32 in.) wide, and are the most complex
    because they handle large tonnages and the cast sections are thick,
    require long cooling and containment sections, extensive secondary
    cooling and complex unbending segments. Many of the problems in con-
    tinuous casting are a direct result of geometric considerations.
    Since most vertical casters have limited headroom, the product must
    be delivered horizontally; therefore, they must both curve and unbend
    the strands. Modern slab casters are designed with a radius greater
    than 10.7 m (35 ft.) and unbending of the slab occurs over several
    roll segments (Figure 3).

Vertical Bow Slab Caster

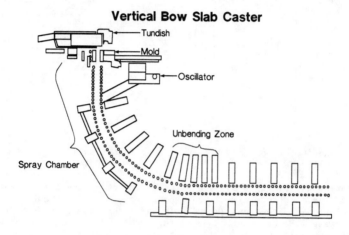

Figure 3

2. Vertical bloom and billet casters - these machines cast shapes
    generally 76 x 76 mm (3 in. x 3 in.) up to 508 x 508 mm (20 in. x 20
    in.) and have similar configuration to slab casters. Although slab
    casters are generally confined to no more than two strands withdraw-
    ing in the same direction, billet casters have been developed to cast
    eight strands continuously. Blooms are usually defined as squares or
    rectangles casting 152 x 152 mm (6 in. x 6 in.) or greater in size:
    billets are smaller, the containment sections for withdrawing the
    strand can be shorter because the shell quickly develops sufficient
    strength to support the smaller sections. The spray chamber is also
    shorter and unbending occurs over a single roll pair. Billet casting
    does not often require submerged pouring and the strand can be cast
    using only oil as a lubricant in the mold (Figure 4).

3. Rotary casters use vertical geometry and centrifugal force to ad-
    vantage by rotating the mold and withdrawal sections as the strand is
    being withdrawn. This rotation produces superior quality round sec-
    tion product because the inclusions are impelled from the periphery
    to be captured by the casting flux. The rotary motion also induces
    refinements in the grain structure and greatly reduces both center-
    line and surface quality problems while maintaining high production
    rates (Figure 5) (4).

Vertical Bow Bloom/Billet Caster

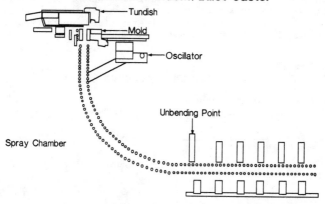

Figure 4

Vertical Rotary Round Caster

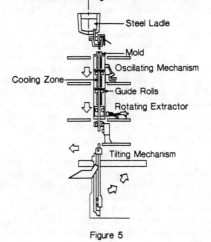

Figure 5

4. Horizontal casting should eventually prove to be an economic method
    for casting specialty steel billets and rounds. Horizontal geome-
    tries greatly reduce investment costs because the structure enclosing
    the machine is not as large. These casters have special requirements
    for modulating the metal flow from the tundish to the mold and employ
    a refractory break ring between the tundish and the mold. Some rely
    on contact heat transfer through graphite molds rather than spray
    cooling for secondary cooling (Figure 6).

Horizontal Continuous Caster

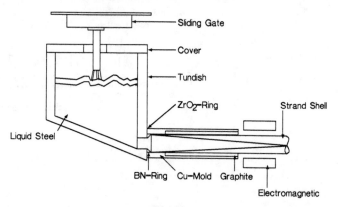

Figure 6

5. Thin slab and thin strip casting - research has accelerated recently
    to evolve innovative casting techniques to cast steel in near-net
    shapes, from 6.35-50.8 mm (1/4 in.-2 in.) thick, which reduce in sec-
    ondary processing of the solidified shapes and eliminate the need for
    roughing stands in the hot strip mill, several longer range programs
    have been funded. A list of active research programs is shown in
    Table 1 which does not include unpublished research programs. Other
    research programs are underway on processes that entirely eliminate
    the hot strip mill by casting strip as thin as 0.76 mm (0.03 in.).
    Some of these cast into the bite of a double roll (Bethlehem, Nippon,
    Nippon-Kokan, etc.); others use a single wheel for solidifying and
    withdrawing the metal (Allegheny-Ludlum, Westinghouse/Armco sponsored
    by D.O.E). The obvious difference between the thin slab and thin
    strip processes is the thickness of the product and cooling rate of
    the product during casting. This may result in marked differences in
    the grain structure and in the morphology and distribution of inclu-
    sions and precipitates in the cast product.

These processes present unique challenges in sensing and control and
may require solutions and approaches not presently available. Even
as the sophistication of the product quality requirements created the
need for special sensing and control features on vertical casters, so
strip casting will further expand the level of sensing and control
required to maintain the integrity of the as-cast product. The need
for this increased level of control is due in as much part to the
elimination of the primary reduction stage with its attendant metallurgical advantages as to the rapid solidification.

Active Novel Steel Casting Research

	Strip		Thin Slab		
	Single Whl.	Twin Whl.	Twin Belt	Horizontal	Other
Allegheny	S/S, 0.50"	X	X	X	X
Bethlehem	X	C, 0.050"	C, 0.3"-1" (USX also)	X	X
Kawasaki	X	Si/SS/C	SS/Si	X	X
Kobe	X	Carbon	X	X	X
Nippon	X	Carbon, Etc.	X	X	X
Sumotomo	X	X	Carbon, Etc.	X	Low Head
Nippon Kokon	X	Carbon, Etc.	X	X	X
Krupp	X	X	Bloom	X	X
SMS—Concast	X	X	X	X	Low Head
Demag	X	X	X	X	Spray
British	X	X	X	Trough	X
West./Armco	Carbon	X	X	X	X
Argonne	X	X	X	X	Moldless Elmg.

Table 1   X—No Reported Research

Modern Slab Caster

The demands on casters for increased productivity and quality have initiated many technological advances. Modern casters are being equipped with automatic systems to assist the operators in the precise control of the machines; the vertical slab caster has received special attention. Figure 7 depicts several of the more germane quality and productivity improvement items (5).

Schematic of Advanced Technologies for Modern Slab Caster

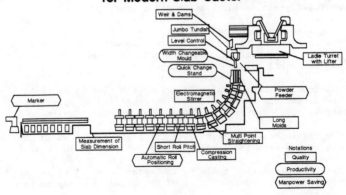

Figure 7

219

Some of these dramatically improve the productivity by adding features which increase the number of casts in a sequence or reduce the turnaround time between sequences. Among these features:

1. Ladle turrets with vertical lift capability to allow transfer ladle shrouding to be removed and replaced without interrupting the cast.

2. Adjustable width molds which allow the strand width to be changed automatically during a cast.

3. Equipment for accurately aligning the mold and first spray chamber segments off-line prior to insertion in the casting.

4. Modifications to the machine design which allow the individual spray segments to be withdrawn along a track or from the side without removing the mold.

5. Inserting the starter bar from the top rather than the bottom enables the starter bar to be inserted before the previous strand leaves the withdrawal section of the machine.

Other features lead to an improvement to slab quality and often improve productivity as well as by allowing the withdrawal speed to be raised:

6. Large tundishes to promote flotation of non-metallic inclusions by increasing the dwell time in the tundish. In conventional strand casting, shrouding of liquid flows between ladle, tundish and mold is important to maintain inclusion-free "clean" steel.

7. Weirs and dams in the tundish to direct the steel flow towards the tundish flux layer in such a way as to trap non-metallic inclusions.

8. Automatic steel level control in the tundish and mold to maintain the liquid metal surface to reduce vortexing of floating slag in the tundish and entrapment of mold powder into the solidifying shell.

9. Electromagnetic stirring in and below the mold to reduce alloy segregation in some grades and promote the formation of favorable equiaxed grain solidification structure.

10. Longer mold section on vertical slab casters to provide longer residence times in the primary cooling area, increasing the entrainment of non-metallic inclusions into the mold flux layer and thickening the shell before exiting the mold.

11. Multi-point straightening to reduce the risk of cracking in the inner radius surface of the slab during unbending by spreading out the high stress zone of strand unbending over a longer length.

12. Compression casting techniques for plate grades which redistribute the stress from the surface to the interior of the slab.

13. More closely spaced rolls in the spray chamber and withdrawal sections to help reduce bulging of the strand and concomitant transverse cracking.

Manual intervention on the casters has been reduced by several other developments:

14. Automatic powder feeding coupled with automatic level control is making automatic casting practical in some steel plants.

15. Automatic monitoring of slab width and thickness, marking of slabs, and improved coordination between the caster and hot strip mill have aided considerably the drive towards process concatenation by increasing the opportunity for hot charging and direct rolling of continuous cast slabs.

These improvements have so increased the quality of cast slabs that ingot product can no longer compete in certain critical grades. An improved understanding of the causes of defects and cracks has made possible the prediction of slab quality on the basis of operational parameters and melt chemistry and temperatures. Statistical Process Control (SPC) models have been developed for many casters which allow the flagging of abnormalities in operation and provide an indication whether conditioning of the slab may be needed. Some of the models contain up to forty variables monitored by the operator or computer during the casting operation and combine to generate a grading for the slab. These models are statistical and emulate past practices but require no new sensors and use only available information. As these models and decision techniques become more sophisticated, they may preclude the need for direct monitoring of slab surface quality off the caster or non-metallic cleanliness of the liquid melt. New sensors are also being developed which could provide more direct information on slab quality. The next few years will provide the experience to resolve the need for direct measurements of slabs or melt quality.

Dynamic Control of Secondary Cooling

Recent developments in vertical slab casting have concentrated on the establishment of dynamic control systems for secondary cooling in the spray segments. In the original systems, the amount of secondary cooling water was directly proportional to the casting speed within threshold limits to prevent the quantity of water from dropping off too far at low casting speeds. The main disadvantage of the method is that the amount of cooling water is increased with increasing casting speed without regard to the surface temperature and no adjustment to the individual spray segments is made when another grade with different water requirements is cast until the first grade has left the spray chamber. Figure 8 shows changes in slab surface temperature in the vicinity of the straightening section and higher in the machine with changes in casting speed. Radical changes in casting speed lead to the formation of transverse surface cracks which must be avoided in most grades.

Dynamic control methods have been developed to maintain constant slab surface temperatures (5). One scheme for this control package is shown in Figure 9. In this method, a surface temperature profile of the strand between the mold and the end of the secondary spray chamber is established for each grade. The process computer tracks the strand and controls the flow rate of spray water to minimize the difference between the calculated and desired slab surface temperatures. The spray segments of the caster are equipped with air-mist cooling which controls the surface temperature of the slab in a tighter range than direct water cooling. The surface temperature exiting the secondary cooling zone is scanned with a two-color optical pyrometer and these data provide feedback to update the control scheme.

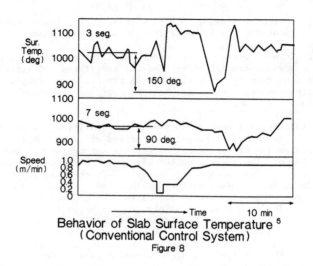

Behavior of Slab Surface Temperature [5]
(Conventional Control System)
Figure 8

Dynamic Control System For Secondary Cooling
After Nippon Steel

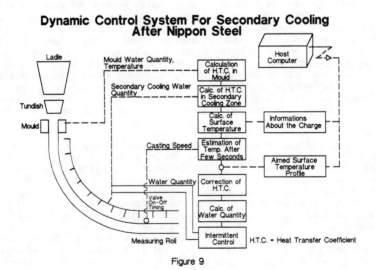

Figure 9

The control schemes make use of a single dimensional conductive heat transfer model and calculate an effective heat transfer coefficient from thermal conductivity data and the measured slab surface temperature at the exit of the spray chamber.

Some results for the slab surface temperature with dynamic control are shown in Figure 10, where it can be seen that dynamic control improves the consistency of in slab surface temperature.

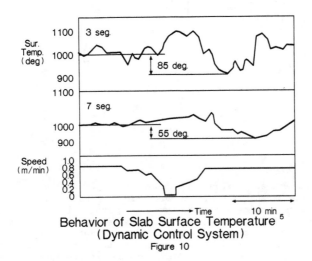

Behavior of Slab Surface Temperature [5]
(Dynamic Control System)
Figure 10

Breakout Prediction Technique

One of the most catastrophic events in the operation of the continuous slab caster is a breakout. Breakouts can occur for several reasons but they always results in a rupture of the steel containment shell below the mold. The molten steel rapidly runs out through the rupture and can destroy roll segments and other elements of the machines. Breakouts frequently occur when the solidifying shell hangs up or "sticks" in the mold as the strand is being withdrawn.

A schematic of the mechanism of "sticker" breakouts is shown in Figure 11. First, an abnormality in the casting process such as a sudden variation in steel level in the mold or inadequate mold powder lubricity, induces a rupture in the developing shell. As the shell is withdrawn, the molten steel being fed from the tundish washes against the healing shell and prevents regrowth. Once the abnormally thin shell exits the mold area, hydrostatic forces overcome the yield strength of the unsupported shell below the mold and the molten metal breaks out.

A method is being developed to detect incipient breakouts and take corrective action before the shell actually ruptures. The method is based on data from embedded thermocouples in the water-cooled mold close to the working face. During normal casting, the shell will develop uniformly as the strand is withdrawn through the oscillating mold. The thermocouples will indicate consistently low temperature readings. If the shell thins due to casting irregularities, the heat flux to the mold increases, and the

223

Schematic of Mechanism for "Sticker" Breakouts in Casters

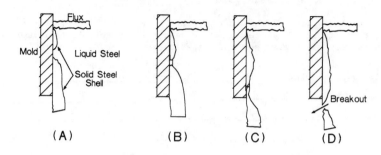

(A)   (B)   (C)   (D)

Figure 11

mold temperature rises. Figure 12 shows typical thermocouple readings for potential breakout conditions. Thermocouple 1, immediately below the metal level, registers a temperature rise resulting from thinning of the shell. A moment later, when the thinning section passes the lower thermocouple, the temperature rise is even greater indicating that healing has not occurred. The breakout prediction logic takes into account the rate of change of the thermocouple temperature readings and the point at which the lower thermocouple reading exceeds the upper thermocouple reading. An alarm is sounded if the model indicates the presence of an incipient break

Typical Temperature & Speed Response During "Sticker" Breakout Condition

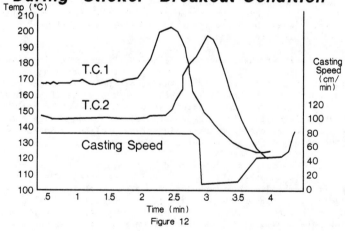

Figure 12

out condition. The operator has 25-40 seconds to reduce the withdrawal speed to allow the shell to thicken before leaving the mold. After the thermocouple readings return to the normal temperature range, withdrawal speed can be increased and normal casting conditions resumed.

Typical positions of the thermocouples are shown in Figure 13. These positions reflect the most likely vertical breakout planes on the basis of experience. The success of the technique for predicting "sticker" breakouts hinges on the hardware methods for attaching the thermocouples, associated electronics to the microprocessor and the logic itself. Continuous refinement is needed to eliminate false alarms which would impact caster productivity.

Schematic of Thermocouple Positions in Slab Caster Mold for Breakout Prediction

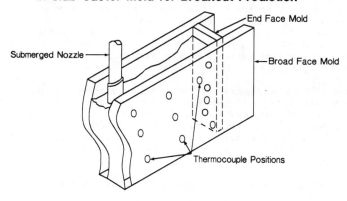

Figure 13

Temperature Distribution Sensors

The American Iron and Steel Institute (AISI) has been active in encouraging collaborative research and development among the steel companies, government R&D laboratories and instrument manufacturers. In 1979 the AISI developed a comprehensive wish list of sensors which would enhance automated processing. This list included some which already existed but required further refinement, several for which the measurement principles had been demonstrated and others for which the need could be stated but the means were open to conjecture.

The common attribute was that none of these sensors was likely to become available without a substantial commitment to research and development. The economic incentives for an instrument manufacturer were somewhat limited - given the high risk potential and the specialized requirements of the end users. On the other hand, the end users themselves, with a siege mentality and operating in a "survival mode" were unlikely to undertake any longer range developments. A strong case was made for collaboration; task groups were established with industry support to promote the development of specific high priority sensors. These groups took an active role in generating the financial support and coordinating the technical effort, and serve to transfer the technology to the industry.

Those priority needs included sensors to characterize porosity in hot materials, to provide rapid in-process chemical analyses, to detect surface defects in strip material, and to determine the internal temperature distribution in solid and solidifying hot bodies.

One of the task groups is working to develop a sensor which will have direct application to continuous casting control problems. The effort is a collaborative development with the National Bureau of Standards (NBS) and Battelle's Pacific Northwest Laboratory (PNL), and is being funded jointly by the Department of Energy (DOE) and AISI member contributions.[6]

For continuous casting processes, a sensing system able to monitor internal temperature distributions provides information which can be used to modulate the distribution and intensity of secondary cooling. This control is needed to avoid excessive thermal and mechanical stresses which can induce surface and internal cracking problems. The approach can also provide information on the location and shape of the final solidification boundary. That information is of critical importance in permitting the machine operator to adjust steel tundish temperature, withdrawal speed, and secondary cooling intensities to optimize productivity. Those critical parameters could also be adjusted to modify the solidification process to retain the unsolidified portions as long as possible to enhance the ability to hot connect cast slabs with the hot rolling process.

The device should operate in real time and non-intrusively, but can make use of ancillary measurements and prior information as available. Approaches which proved feasible include eddy currents which depend on electrical resistivity and magnetic permeability, and ultrasound which relies on elastic modulus and density. The eddy current approach is well-suited to aluminum and may also prove workable on steel products above the Curie temperature.

Ultrasonics is the method of choice at NBS (6), where the work has included:

- the refinement of calibration data on the variation of ultrasonic velocity with temperature in various materials;

- the experimental measurements of ultrasonic time-of-flight in cylindrical and rectangular samples up to 700°C;

- the development of algorithms to reconstruct temperature distributions from such information.

The progress on this sensing system has reached the point where the only impediment to the development of a practical instrument seems to be the temperature tolerance of the EMAT (Electro Magnetic Acoustic Transducer) device.

The D.O.E. is sponsoring efforts at PNL to develop appropriate high temperature EMATs, emphasizing the engineering aspects with the intention of demonstrating a practical device in a steel plant environment.

In early November, 1986, an EMAT sensing system capable of continuous operation on heated steel samples was tested successfully on 23 cm lengths of 304 stainless steel bars heated to 1100-1500°C. This system continuously monitored the time of flight of ultrasonic pulses injected by means of a pulsed 4J ruby laser (25 nanosecond pulse duration) as the sample was heated from room temperature to above 1100°C.

Lift-off distances of up to one centimeter were found to provide adequate sensitivity on room temperature samples. Trials are underway to extend sample temperatures beyond 1300° C, and to demonstrate the system on high carbon steel billets to comparable temperature ranges. The next step is to install a prototype system in a laboratory scale casting machine. This approach will test the system under conditions close to that of an operating facility, and allow extended testing on materials with both solid and solidifying centers. This testing will provide a great deal of information and is regarded as an important first step towards commercialization.

Several system requirements must be addressed including the method of generating the initial ultrasonic pulse - the standards work at NBS uses a LASER. We would prefer to avoid the complexities and cost of the LASER system and are proposing to use the same EMAT technology to both initiate and receive the pulses. If that proves out in laboratory testing at PNL the immediate temperature sensing problem will be solved and a new class of sensing devices which can solve innumerable problems, including several others on the AISI sensor needs list will have been developed. It should be understood that the EMAT approach is sensitive to a number of metallurgically important parameters. That sensitivity presents both problems and opportunities for in-process characterization. This EMAT device becomes a generic sensor capable of solving problems other than the one it was designed to address.

A second requirement is more subtle and has to do with the need to measure the geometric path length of the ultrasonic pulse used to interrogate the material. NBS is addressing this problem through special approaches to the reconstruction algorithms which infer temperature fields from ultrasonic information.

Future Technical Needs

Even with all of the developments of the last fifteen years, improvements in several other technologies would greatly benefit efficiency of plant operation and product quality.

One example of a technology that is needed is a method to limit slag carry-over from ladle to tundish at the end of a cast. This carry-over slag is very harmful to refractories in the tundish and shroud and can be entrained by vortexing into the mold. It appear as inclusions in the slab (Figure 14) (7). To date, no cost effective method has been found to control this slag carry-over. On the basis of water modeling, we have determined that vortexing can occur when the metal level in the tundish drops below 60 cm (23.6 in.) with tundish slide gate control or 50 cm (19.7 in.) with stopper control at Armco's Middletown slab caster with average speed 84 cm/min (Figure 15) (8). On Figure 15a, speeds to the right of the lines indicate conditions for vortexing and the lower speeds to the left indicate no vortexing. For most slab casters casting common grades, build-up of slag in the tundish is the principal reason for limiting the number of heats through the tundish.

Aggressive research into innovative new casting will eventually lead to practical strip or thin slab casting technologies. Without question, however, these methods will require even more sophisticated and complex sensor and control techniques than have been developed to date for present casters. Although techniques in these fields are still a matter of

Tundish System Showing Conditions of Slag Vortexing Into Mold

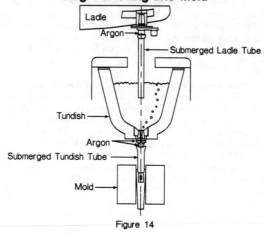

Figure 14

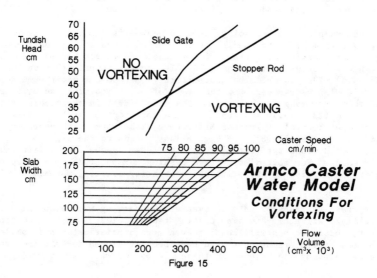

Tundish Head cm

Slide Gate

NO VORTEXING

Stopper Rod

VORTEXING

Caster Speed cm/min

Slab Width cm

75 80 85 90 95 100

Armco Caster Water Model

Conditions For Vortexing

Flow Volume (cm³x 10³)

Figure 15

228

conjecture, some sensor and other technological needs may be anticipated:
a) laser devices for on-line measurements of strip width and thickness, b)
advanced sensors for both heat flux and temperature readings through difficult environments, and c) techniques for closely controlled highly localized heating of refractories by plasma, resistance or induction heating.

Special handling methods will be needed to collect strip at speed, and
techniques will be needed to control the process by means of heat flow and
temperature algorithms and special motor controls required for delicate
speed control. Electro magnetics may play a role in metal-feeding nozzles,
strip hold-down methods or even metal feeding to the holding vessels.

Finally, after 100 years, we are beginning to come to grips with some
of the problems that must have been in the back of Henry Bessemer's mind
when he first proposed continuous casting. Good progress has been made
during the past fifteen years. The next 100 years may prove to be the gol-
den age of continuous casting.

References

(1) T. Harabuchi, Continuous Casting of Steel - Design, Operation and
     Maintenance, U. of Michigan Summer Conference, May 14, 1984.

(2) American Iron and Steel Institute reports.

(3) Office of Technology Assessment, Washington,D.C. "Benefits of
     Increased Use of Continuous Casting by the U. S. Steel Industry" Oc-
     tober, 1979.

(4) Industrial Heating, "Rotary Continuous Casting, Enhances Control of
     Crystallization in Producing Tube Rounds", September, 1982.

(5) H. Iso, M. Tezuka and D. Mitukoshi, "Continuous Slab Casting at
     Kimitsu Works of Nippon Steel Corporation". Presented in China, De-
     cember, 1985.

(6) F. A. Mauer, H. Wadley, S. Norton, R. Heinrick, J. R. Cook, "An
     Ultrasonic Method of Determining Temperature Distributions in Hot
     Metal Bodies," Instrument Society of America Conference, October 14,
     1986, Houston, Texas.

(7) P. G. Boting, P. J. Kreijer, A. van der Velden and J. H. Verhens,
     "Considerations on Productivity and Quality of the Slab Caster at Ij-
     muider", International Iron and Steel Congress, London, England, May,
     1982.

(8) D. Follstaedt, Internal Armco Report.

Methods for economic use

of material and energy in ring rolling

P. Dreinhoff and R. Kopp

Institut fuer Bildsame Formgebung
Rheinisch - Westfaelische Technische Hochschule
Aachen, W.-Germany

Abstract

When rolling rings with rectangular cross sections, one often observes faults like contractions, dishing, and out-of-roundness of the finished rolled ring. This is a reason for extensive material addition to the finished geometry of the ring. Reducing this additional material would mean considerable saving of raw material and energy. A rolling strategy has been developed considering the causes of shape faults. It describes the ratio of reduction in the radial and axial gap depending on the continuously changing geometry of the rolled ring. Faults are avoided by using these results in the rolling process. With the application of adaptive process models the rolling process can be improved with respect to the performance limits of the mill (forces, torques). In order to calculate the control variables with sufficient accuracy, the force and torque models are continuously updated with measured values of the present process.

As a result of this research and development there has been a considerable reduction of material (raw material and energy saving up to 20 %). Also, intermediate heating may become unnecessary by minimizing rolling times because of optimal use of the mill's capacity. The geometric spectrum of rings has been enlarged allowing more extreme ratios of thickness to height of rings.

Introduction

Research on the subject of ring rolling began about 5 years ago at the Institute of Metal Forming at the Technical University of Aachen, W.-Germany.

The acquisition of a ring rolling mill of industrial size financed by public and industrial funds allows empirical examination of this rolling process under standard industrial conditions.

Ring rolling has been in practice for a long time, but until now few studies of the fundamental principles of the radial-axial ring rolling process has been published. Initial researches have been concerned with studying the fundamentals of the rolling process, e.g., the geometry and the kinetics as well as the cause of faults of rings with square sections (1). What is really missing is a process control based on the forming conditions, because this has been done mostly by experience.

A computer program for an optimal process control based on the results of findings of this research work has been realized. Another task was to make use of the capacity of the mill, that is the maximum rolling force, torque and growth rate. Therefore adaptive models have been developed. The basic thoughts and the concept of this adaptive process control based on the new rolling strategy for square cross sections are the topic of the following.

Faults of rolled rings

Some typical faults in ring rolling cause additional material and oversizing the ring geometry compared to the final machined ring geometry (Fig.1).

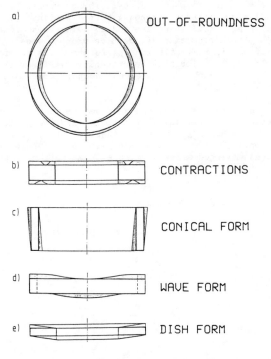

Fig. 1: Typical faults of rolled rings

232

The first fault shown in this figure is the 'out of roundness'; this is often to be found when rolling tubular rings (Fig. 1a).

The second fault is concerning the rectangular shape of the ring. These faults, contractions on the flat surface of a disc-shaped ring as shown in Fig. 1b, and contractions on the circumference of tubular rings, are discussed later. The formation of these contractions is based on the forming geometry in the two roll gaps.

Further, there can be a conical and a wave form of the ring, which can often be avoided by an exact installation of the ring mill (Fig. 1c).

The last fault shown in Fig. 1e is the dish form of a ring that depends on the forming conditions in the roll gaps, similar to the fault in Fig. 1b.

The sum of all these faults causes an extensive material addition. The measured percentages of this extra material influenced by these special faults are shown in Fig. 2. The data of this diagram have been derived from measurements of industrial rolled rings with square cross sections rolled by different ring-rolling mills. The sum of the single faults is the total mean value for the additional material that varies according to the outer diameter of the ring.

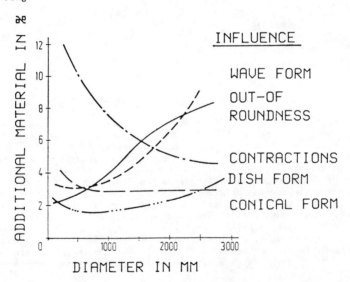

Fig. 2: Additional material influenced by different faults

Avoiding these faults, reducing the overmeasure and saving material even means a great saving of energy. Due to some geometry forms up to 50 % of energy can be saved (2). On rolling profiled rings this savings can be greater.

<u>Ring rolling process</u>

The difference between the ring rolling process and other known rolling processes, such as sheet rolling, is its non-stationary behavior. A non-stationary process is characterized by the fact, that the mean state variables vary. Such a process is also called time variant since the mean state variables are functions of time. This non-stationary behavior is caused by the continuous rolling of the ring in two rolling gaps: the radial and the axial gap (Fig. 3). The ring is formed radially by advancing the mandrel roll towards the main roll, and axially by lowering the upper conical roll.

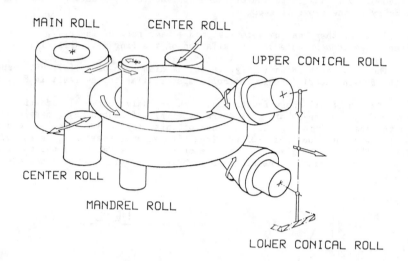

MAIN ROLL   CENTER ROLL

UPPER CONICAL ROLL

CENTER ROLL

MANDREL ROLL

LOWER CONICAL ROLL

Fig. 3: Principle of radial-axial ring-rolling

Due to the reduction of the ring cross section, the diameter grows and the ring changes its position in the axial gap. The rolling gaps interact by causing axial and radial spread, thereby increasing the reduction in the respective opposite gap. The following geometric quantities are affected by the non-stationary behavior of the process: The ring cross sections are continually reduced in the rolling gaps, so that the forming zone geometry changes even for pre-set deformation rates. The contact lengths between the rolls and the ring in the radial gap changes because of the decreasing ring curvature. The reductions at the main and the mandrel roll change for constant total reduction in the gap. The working radius of the conical roll grows with increasing ring diameter in slip free rolling being accompanied by an increase in the contact area when the reduction is held constant. The period of the ring increases according to the growing ring cicumference.

Ring geometry during rolling

A snapshot of the ring during the rolling process, Fig. 4, shows that the wall thickness exiting the radial gap would increase from s_1 to s_0' because of the reduction of the rolling gap during the last revolution, if the wall thickness would not have been changed in the axial gap. But, in reducing the ring height, a radial spread appears, increasing the wall thickness entering the radial gap. In Fig. 4, the rolling gaps are represented by discontinuities, the spread by its mean over the ring height. The total reduction in the radial gap results from the addition of the mandrel roll advancement of one ring revolution and the spread in the axial gap. Corresponding conditions exist for the entering and exiting ring height in the axial gap. The spread in both rolling gaps is described by an empirical

234

equation especially developed and adapted for the rolling process (3).

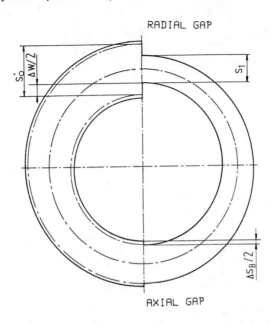

Fig. 4: Surface view on a ring during the rolling process

The reason for the contractions mentioned above is that in the rolling process only small deformations up to 10 % can be carried out in the two roll gaps.

When rolling a disc-like ring, only the outer and the inner areas in the radial gap are deformed by compressive stress. The area inbetween remains nearly rigid and prevents a stretching of the outer and inner areas.

The result is that the material flow is not axial. The ring spreads in the outer and the inner cicumference. A swelling occurs in the plastic zones at the main and at the mandrel roll. This irregular cross section enters the axial rolling gap half a rotation later and because of the shape it is subject to an irregular axial height reduction.

These reductions mean an unequal extension of the ring cross section. They are larger at the outer and inner cicumference than in the middle zone (Fig. 5).

Because of the inner coherence of the material, the ring cross section is stretched equally in its entirety, which means that the middle zone is extended in a greater degree and must therefore contract. This depends on the difference of the reductions in the roll gaps. For this reason the radial deformation, which causes swelling, must be matched to the axial deformation, so that contraction is not possible. For rolling a tubular ring the relationship is reversed.

Following these requirements during rolling process, the overmeasuring of material, as well as the effort of machining can be considerably reduced.

235

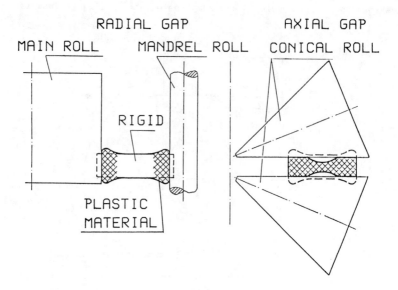

RADIAL GAP

MAIN ROLL   MANDREL ROLL

RIGID

PLASTIC
MATERIAL

AXIAL GAP

CONICAL ROLL

Fig. 5: Forming conditions in the rolling gaps

Since the ring geometry continually changes during the rolling process, the deformation and therefore the tool feed rate must always be matched to the instantaneous conditions.

The momentary geometric conditions can be calculated by the computer. Now there has to be a criterion for the control of the rolling process.

This criterion is one of the results of the research work, the so called rolling gap ratios that are decisive concerning the occurrance of contractions of the rings cross section (3).

This factor is important during forging or conventional rolling too. It corresponds to the bite ratio in forging. Empirical researches showed that a rolling strategy could be formulated that requires an harmonized rolling gap ratio of the radial and the axial gap depending on the geometry of the tool, that means the diameter of main and mandrel roll and the working diameter of the conical rolls, and even the instantaneous geometry of the ring.

The reduction in the radial and axial rolling gaps are coordinated in such a manner that the above described faults do not appear . The ratio of the reductions depends on the ring type (disc, tubular ring) and the momentary ring geometry which varies during the rolling process. It is self evident that the blank shape must have the required shape for this strategy to be applied.

The rolling strategy for faultless ring cross sections requires only a fixed ratio of radial to axial deformation, not the absolute values. This degree of freedom allows the rolling process to be adjusted to the capacity of the mill, and with that, reducing the required rolling times, avoiding reheatings, and reducing the amount of energy and raw material needed.

The process models must be matched to the actual process occurrences in
order to achieve agreement between the calculated results and the actually
measured values.

First of all, a formula for roll forces is formulated from the physical
relationships, which enables an approximate calculation of the rolling force.
Naturally, the yield stress is included in this model.

The material behavior is non-stationary, just as the ring geometry. It
can be described by the mean yield stress k_{fm} as a function of the deformation, the structural condition and the material temperature. So, the mean
yield stress depends on the already mentioned non-stationary forming zone
geometry. The variation of forming times and pause times (the pause times
increase with growing diameter) causes non-stationary strain hardening and
removal of work hardening. A change in the temperature distribution follows,
the temperature being a quantity with a strong influence on the yield stress.
The calculation of the rolling forces and momenta is based on the elementary
relations: the rolling force is the product of deformation resistance and
area of contact, while the momentum is derived from the product of rolling
force and moment arm. Thus, one must know the yield stress and forming
efficiency. In ring rolling, the forming efficiency mainly depends on the
shear stress caused by the relatively small momentary deformation in the
rolling gaps. Thus, it can be described as a function of the forming zone
geometry.

The rolling forces cannot be calculated accurately enough if the yield
stress is taken from data collections. Experience shows that their accuracy
is questionable, especially because the individual forming histories are not
sufficiently taken into consideration. Furthermore, the most important quantity influencing the yield stress, the representative temperature in the
forming zone, cannot be measured with the required accuracy.

Calculation model for the non-stationary process

All of these problems can be solved by taking the needed information
directly from the running process.

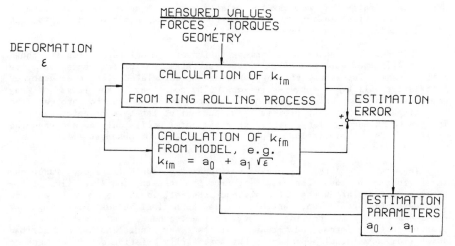

Fig. 6: Estimation of model-parameter

237

A simple empirical relation is formulated for the yield stress k_{fm} as a function of the deformation ϵ, with two parameters a_0 and a_1, still to be determined by experiment. They are adapted to the actual process as follows, Fig 6: the momentary yield stress is calculated from the rolling force, the area of contact and forging efficiency, which depends on the geometry. The yield stress can also be calculated by way of the model equation with the deformation rate. The model must be corrected if there is a difference between these two values by readjusting the parameters a_0 and a_1. This adaptation (parameter adjustment) results from the requirement, that the square mean of the difference between the values resulting from the model equation and the measured values becomes minimal (4), e.g. the least square method. Using this algorithm during the rolling process allows the correction of the parameters in such a manner that the process characteristics become better calculable and the model is able to follow the non-stationary behavior. The following influences make the continuous estimation of the model-parameters necessary:

- material history (forming and temperature history)

- simplifications in the model equations

- simplifications in the calculation of the forming efficiency

- unknown boundry conditions such as state value of the mean ring

  temperature, spread and friction.

In contrast to a stationary process, it cannot be expected that the model parameters become constant after a certain amount of process quantities have been measured and subsequently adapted with the process model. In stationary rolling processes such as sheet rolling (the stationary state is reached shortly after commencing the process), final parameter values set in after a certain amount of time. In non-stationary processes, no constant parameter values set in; the parameters vary in time. The model-equation may only be applied in time intervals which can be considered stationary, that is in which the conditions (for one revolution of the ring, for example) don't change markedly. The latest measured values must be mainly used for the estimation due to the constantly changing process characteristics. This is done sufficiently by reducing the weighting of past measured values with a weighting-function in a recurrent parameter estimation algorithm. This can also be described as a "subsiding memory".

The result of an on-line implementation of a model-adaptation is shown in Fig. 7, the quotient of calculated and measured rolling force vs. the rolling time. The actual rolling force can be calculated with an accuracy of $\pm$ 5% after only a few ring revolutions. The model has no knowledge of any material characteristics, so that the parameters are set to zero at the beginning of each rolling process, independent of the type of material used. It can be shown that large differences and model faults only appear when the limits of the "quasi-stationary process state" are not observed (5).

The process control which uses this adaptive models (Fig. 8), functions as follows: the equations resulting from the geometry and kinematics of the rolling process allow the calculation of the roll advancing velocities for the next ring revolution, which are taken as nominal values fulfilling the requirements of the rolling strategy on the one hand, and leading to the planned rolling forces and torques, on the other hand. The above described adaptation of material behavior to the momentarily existing conditions ensure sufficient accuracy in the precalculation, since the process is quasi-stationary for one ring revolution.

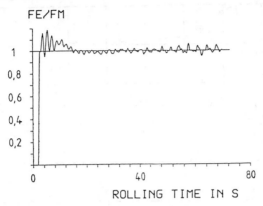

Fig. 7: Ratio of the estimated rolling force (FE) and measured rolling force (FM) vs. the rolling time

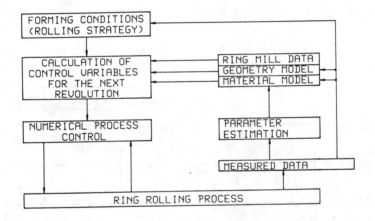

Fig. 8: Concept of the process control

The following diagrams show some examples of rolled rings with this new control algorithm. The first diagram (Fig. 9) demonstrates the radial force vs. the rolling time. The limitation of the force was set to 500 kN. In the beginning, after a short over-swing, the force is limited to the preset value with good accuracy by calculating the deformation rates one revolution in future. The over-swing in the first seconds is produced by the noncircularity of the forged blank ring.

The next diagram (Fig. 10) shows the limitation of the radial torque which was limited to 40 knm. After a short over-swinging in the starting-phase of the rolling process the constant preset value is reached. Further on, the ring is rolled with constant radial torque.

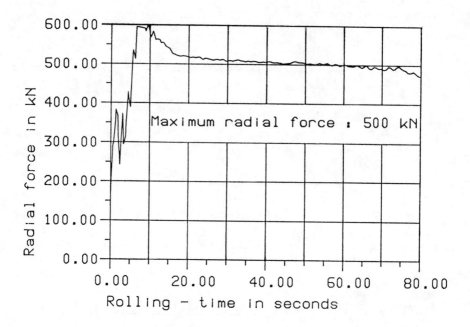

Fig. 9: Radial rolling force vs. rolling time

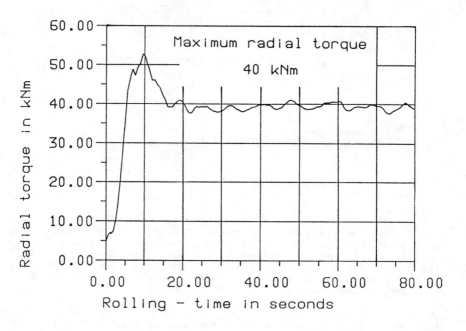

Fig. 10: Radial torque vs. rolling time

The utilization of the mill capacity is secured even for unknown material behavior. The calculated parameter values of the equation for the mean yield stress can be stored and used for an improved and more accurate precalculation of the process.

Summary

The advantages of this process control as opposed to the existing possibilities in the ring rolling industry can be summarized as follows:

The additional material can be reduced because it is possible to roll faultless rings with rectangular cross sections without contractions. Because of this new rolling strategy and the exact position control of the rolling gaps the ratio of thickness to height of the ring is now extended to 8:1, so that extreme flat disc-like rings can be rolled.

The forces and torques of the rolling process can be calculated in future with good accuracy, so that the full capacity of the mill can be used. This means shorter rolling times. It is not necessary to know the exact material behavior, because this is estimated by adaptive process models.

This new computer control simplifies the operation of rolling rings with rectangular cross sections. Pilot tests for a new ring geometry or unknown material behavior are not necessary.

Acknowledgements

This project was commendably financed by the AIF, the society of the industrial research foundations in Germany, by the DFG, the german science foundation, the ring-rolling industry and the State of Nordrhein-Westfalen.

REFERENCES

1. Koppers, U.: <u>Geometrie, Kinematik und Statik beim Walzen von Ringen mit Rechteckquerschnitten</u>, (Ph.D. thesis, RWTH Aachen, 1987)

2. Koppers, U. et al.: Methoden zur Verringerung des Material- und Energieeinsatzes beim Ringwalzen, (<u>Stahl und Eisen 106</u>, Nr. 14/15, 1986)

3. Koppers, U.: Abschlußbericht zum Vorhaben: <u>Untersuchung zu den Grundlagen des Ringwalzens und Weiterentwicklung des Verfahrens</u>, (Ministerium für Wirtschaft, Mittelstand und Technologie des Landes NRW, 1986)

4. Isermann, R.: <u>Prozeßidentifikation</u> - Identifikation und Parameterschätzung dynamischer Prozesse mit diskreten Signalen, (Springer Verlag, 1974)

5. Dreinhoff, P.: <u>Entwicklung eines Prozeßmodelles für das Walzen von Ringen mit Rechteckquerschnitten</u>, (Abschlußbericht zum AIF Vorhaben, AIF Nr. 5709, 6170, 6720, 1986)

6. Dreinhoff, P. et al.: Adaptive Prozeßführung für das Ringwalzen, (<u>Stahl und Eisen 106</u>, Nr. 10, 1986)

242

MODELING, SENSING AND IN-PROCESS CONTROL OF

SHEET DEFORMATION PROCESSES

David E. Hardt

Laboratory for Manufacturing and Productivity
Massachusetts Institute of Technology
Cambridge, Massachusetts 02139

ABSTRACT

The control of sheet deformation processes has received growing attention in the industries most dependent upon this process, in particular the aerospace and automotive. Most of this attention has been placed on classical areas of engineering science such as applied mechanics, plasticity, and materials science, with an emphasis on constitutive relations for high strains, new materials, and numerical analyses of the basic process mechanics. However, the basic uncertainties inherent in these processes arising from both material property and processing environment variations will continue to confound even the most accurate and detailed analyses. In addition the geometric complexity that characterizes manufacturing products similarly confounds numerically based approaches.

In this paper this problem is addressed from the point of view of in-process material *and* process property identification and control. Using the concept of a "control model", that is one derived from basic physics, but simplified to suit the real-time control objective, is forwarded. This model is then used to specify the necessary measurements to implement control of the process. This in turn leads to a process control system that is particularly robust to material and processing variations and can often eliminate sensitivity of the outcome to such influences. These concepts are illustrated through three examples and cases studies: brakeforming, roll bending and 3-D stamping, and each demonstrates different approaches within the same methodological framework.

INTRODUCTION

The forming of sheet metal to complex two and three dimensional shapes represents one of the most material and energy efficient ways of manufacturing mechanical parts. However, the current practice in most industries is to use highly iterative forming procedures (for low lot size parts) or to employ extremely costly fixed configuration tooling. In either case, costly and time consuming setup procedures are required, and the formed product inevitably has a high variance in dimensions.

The basic cause of these problems is found in the nature of a sheet deformation process. The plastic deformation involved becomes a strong function of the intrinsic properties of the material (such as yield point and flow mechanics), the geometric properties (such as thickness and sheet size), and the interface conditions (such as tool-sheet friction). Because of this uncertainty, methods that attempt to predict sheet behavior, no matter how detailed in their derivation, must always contend with a high degree of uncertainty in calibration.

Accordingly, we have pursued a novel approach to the problem of controlling sheet deformation processes that calls upon the basic principles of control system theory, and makes use of in-process measurement, control modelling and real-time control algorithms to achieve a more consistent, material property insensitive process. In this paper, the basic elements of the approach are outlined, and through the use of three examples, the various manifestations of this methodology are presented.

CLOSED-LOOP PROCESS CONTROL

To distinguish the type of process control presented here from conventional manufacturing automation, it is only necessary to distinguish between *machine control* and *process control*. In the former, a closed-loop control system will be installed to insure that the processing machine (such as a forming press) has highly repeatable performance. For the case of a forming process this may involve very tight control of the tooling displacements or the forming forces. Figure 1a illustrates this type of control (which is often denoted as Numerical Control or NC) and from this diagram it is clear that the workpiece is excluded from the control loop, thus the system cannot react to any unexpected behavior in the workpiece. However, as we have discussed above, for the process of sheet forming the workpiece behavior is inherently uncertain.

To solve this problem, at least conceptually, we need only extend the control loop to include the workpiece, as is shown in Fig. 1b. Here the reaction of the workpiece to the processing machine is measured directly and the control system now seeks to produce the correct *part* rather than simply concerning itself with correct *machine* operation. To effect this or any type of closed-loop control, there are three essential elements required:

1) A causal or **control model** of the process
2) A set of **measurements** made in-process
3) An appropriate **control method** based on 1) and 2)

In what follows, these concepts will be illustrated through three different processes, each presenting different opportunities for applying this concept. For each the three elements necessary for control will be discussed.

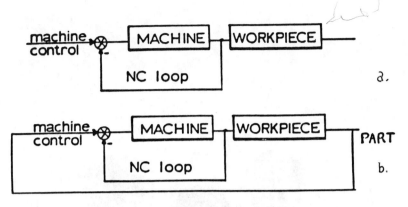

Figure 1 Manufacturing Control Systems
a) Machine or Numerical Control (NC)
b) Process or Part Control

The three processes are: simple three point bending (brakeforming), continuous two-dimensional roll bending, and general three-dimensional sheet
stamping. In the case of brakeforming (illustrated schematically in Fig 2)
the goal is to bend the material such that when the load is removed, it will
"springback" to the correct included angle for the part. In the case of roll
bending (Fig. 3) this local bending is spread out continuously along the
workpiece, and appropriate control of the center roll position must be
applied to correctly overbend each point along the workpiece. In the case of
three dimensional stamping (Fig. 4) the problem is one of correctly designing
the matched die and punch set to compensate for in-plane as well as bending
springback. (We will not consider the equally important problem of material
flow control in 3-D forming, although it has also been addressed as a control
problem (1).)

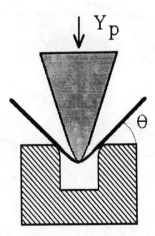

Figure 2 Geometry of a Simple Brakeforming Process

245

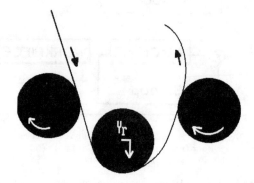

Figure 3 Roll Bending using "Pyramid" Rolls

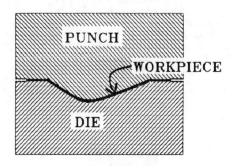

Figure 4   Schematic Cross-Section of Matched Die Stamping
(Note that edge constraints have not been shown)

In what follows, the modeling, measurement, and control aspects of each of these process will be discussed, and these three processes will illustrate the various ways in which real-time control can impact sheet deformation.

CONTROL MODELS OF DEFORMATION

BRAKEFORMING

Since it involves a nearly ideal mechanical situation, brakeforming is one of the simplest forming conditions to analyze. The basic geometry shown in Fig. 2, illustrates the control inputs (the punch displacement Y_p, and the final output, the unloaded angle Θ_u. The basic problem is to relate these two, and this relationship is governed by the moment curvature (M-K) relationship for the sheet undergoing bending. Thus an appropriate control model is one that develops this curve during the process, and then relates it to the sheet angle. A typical M-K curve is shown in Fig. 5. Once this basic constitutive relationship is known *for the workpiece that is in-process*, it must then be applied to all points along the loaded section of the sheet. This model then dictates the necessary measurements for this process, which will be discussed in the next section.

ROLL BENDING

The basic geometry of roll bending is shown in Fig. 3. Again the sheet is subjected to three point bending, but here the material is continuously rolled through the press, and each point on the workpiece can see a different bending moment and can, therefore, take on a different permanent curvature. Thus, this process can produce two-dimensional parts with continuously varying curvature along the arc-length of the part. Again the key here is to control the springback of the part, since the process input (center roll position Y_r) and the final curvature relationship are quite complex and unknown *apriori* because of both the complex elastic-plastic geometry of the piece as it is rolled through, and because of the springback. However, the local control of each point along the piece that this process affords leads to a different and ultimately simpler control model.

If we refer to the M-K diagram of Fig. 5, and realize that this diagram also represents the forming history a single point on the workpiece as it enters, reaches the maximum moment point under the center roll, and then leaves the press, we can see that the key is to control the maximum moment and curvature (M_{max} and K_{max}) conditions such that the unloaded curvature (K_u) will be correct. This can be described by the simple relationship:

$$K_u = K_{max} - M_{max} / S \qquad (1)$$

where S is the slope of the elastic loading (and unloading) line on the M-K diagram.

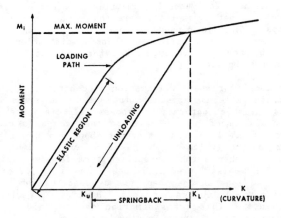

Figure 5   A Typical Moment Curvature Relationship

While this model is sufficient to control the process in a quasi-static fashion (2), when faster operation is required, a more complete model is necessary. Such a model was derived by Hale and Hardt (3), and it has been shown to correctly predict the basic process and process stability characteristics under high speed control. This model, the essence of which is illustrated in Fig. 6, shows that when center roll position is related to the unloaded curvature, the net control model looks like a deadband process with an uncertain deadband width. This width is a function of both the yield point of the material and the speed of rolling during forming.

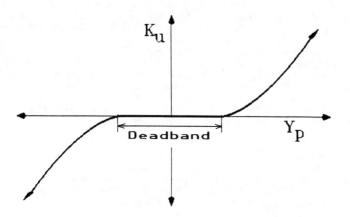

Figure 6 The Input-Output Relationship for Roll Bending

THREE DIMENSIONAL FORMING

In the case of 3-D forming, the typical analytical approach is to attempt to completely capture the mechanics of the elastic and plastic deformation processes by using deformation theory (if the geometry is simple) or finite element numerical methods for general geometries. In either case the objective is to determine sheet formability more than to predict actual tooling-part relationships. This is caused, in part, by the complexity and computational burden imposed by specific part shape specifications, and also, as mentioned above, by the difficulty in knowing the material and process properties for each workpiece.

Accordingly, our model for sheet deformation concentrates on the integrated effect of the strain patterns imposed on the sheet: the shape. However, if one simply models the sheet deformation process as a die shape-to-part shape relationship, the choice of how to describe that shape is central to the model formulation. Since local changes in die shape can have a global effect on the sheet deformation (which is obvious from even a simplified membrane model of deformation) then a the appropriate form for describing shapes is one that captures this distributed influence. As is discussed in (4), one way to do this is to look at the shape not as described by a 2-D array of coordinates (e.g. $z(x,y)$), but rather as a 2-D array of equivalent spatial frequency components. This transformation can be readily accomplished through application of the Discrete Fourier Transform (DFT). By using this transformation (which essentially fits a sine wave series to the surface contour), we not only capture the influence of local displacement changes on global part shape, we now have a description of shape in a form that is compatible with linear system theory.

Thus if we are given a die shape (the process input) and the resulting part shape (the process output), and each is described by the DFT of the 3-D coordinates : D(u,v) and P(u,v) respectively, then we can, by definition, define a Deformation Transfer Function as the ratio of the output to the input:

$$H(u,v) = \frac{P(u,v)}{D(u,v)} \qquad (2)$$

This transfer function now illustrates how an input (the die shape) is transformed into the output (the part shape). While it does not in any way predict the details of the mechanics of the deformation, it can capture the essential features of springback, since springback is a linear elastic phenomenon. This transfer function serves as a linear predictor of the workpiece behavior, it can be used directly to develop a control system, as will be detailed below.

IN-PROCESS MEASUREMENT

The availability and robustness of measurements for real-time control is the key to success of this concept. However, most of the desirable measurements (the final part shape in-process) are not available, so an approach that examines feasible measurements in the light of the above control models has been taken. Since all of these processes, and the models that are used for control, are essentially concerned with mechanical behavior of the sheet and the process, the basic measurement of force and displacement are dominant. Again, however, the individual processes have different needs and opportunities for such measurements.

BRAKEFORMING

While it would be desirable to directly measure the angle to which the sheet is formed, this often represents an unacceptable addition to the tooling complexity, and its robustness in a production environment is questionable. However, it is quite straightforward to measure the forming force and punch displacement (Y_p in Fig. 2) on most machines (see for example (5)). The trade-off between measuring angle versus displacement relates to the amount of sheet geometry that must be modeled in the control system. As detailed above, the brakeforming model is essentially one of describing the moment curvature relationship and then applying it to present loaded state of the sheet. If the angle is not measured, then it must be inferred from the integral:

$$\Theta = \int_0^{s_{tot}} K(s) \, ds \qquad (3)$$

where K(s) is a known curvature distribution, Θ is the total angle and s_{tot} is the total arc length of the deformed section. Thus we can see that if the angle is not measured, the curvature distribution K(s) must be known. As pointed out in (5), it is precisely uncertainty in K(s) that leads to errors in the final control of such a system. Thus the trade-off in brakeforming is between simple, rugged measurements that entail some inherent uncertainties, and more complex measurements that obviate error producing calculations.

ROLL BENDING

In this case the measurements are more direct, but again a trade-off between the most direct measurement and the most reliable exists. The model forwarded above requires the maximum moment and curvature of the workpiece at the center roll. The former can be measured by instrumenting one of the outer rolls for force, and is readily accomplished. However, the direct measurement of curvature at a point is impossible and compromises must be made. In (2),(3), and (6), several approaches are detailed, however, they all can be reduced a single method that measures the local change in angle as the workpiece moves past the center roll. The main difference with each approach is the interval over which the change is measured. If a high degree of spatial resolution is required (i.e. to measure rapidly changing curvature), then the interval must be small any local workpiece displacements relative to the center roll must be sensed. However, as this interval in increased, the signal-to-noise ratio of the sensor improves since large displacements result for the same given angle change, but this is accompanied by a loss of spatial resolution. In the limit as this distance become the spacing of the outer rolls (see Fig. 3) and the measurement reduces to the center roll displacement Y_r. Since this can be very easily and accurately measured, it greatly simplifies implementation of the control system, and as shown in (3) this method remains very accurate for small palstic deformation of the section (such as involved in straightening operations).

THREE DIMENSIONAL FORMING

Again it is useful to consider measuring generalized forces and displacements for 3-D forming, but in this case only displacements are relevant. This is because in most cases of sheet stamping the sheet deformation modes involve both bending and in-plane tension or compression. Furthermore, the deformation is not at all homogenous or monotonic as with pure bending, so local forces or stresses must be sensed to be of use. Therefore, appropriate sensing would include the measure of both normal pressures everywhere on the tooling surfaces *and* measurement of local surface shear stresses. While both measurements can be made, to consider a complex set of tooling with a large number of transducers would be ludicrous.

Accordingly, it must be conceded that to be practical, only sheet displacement measurements can be made, and these will take the form of three dimensional coordinates on the surface of the part. Furthermore, such a measurement can only be made after the part has been formed, since the deformed part shape is known, or expected to be the same as the tooling shape. Such coordinates can either be measured by conventional coordinate measuring machines (which are very accurate, but cumbersome and slow) or by optical methods, such as those using structure light and computer video systems (which are quick but lack high accuracy). In either case, in-process measurement remains a distinct problem in trying to both rapidly and accurately determine sheet contours in-process.

CONTROLLER DESIGN

Once a set of control models and associated measurements are established, it is possible to develop specific control system designs for each process. As shown in Fig. 1, this generally involves working with a measured error between the actual and target output for the process, and a process machine input that is commanded on the basis of this error and its recent history. In the three examples below, three different methods are illustrated. For roll bending the approach is identical to classical feedback or servo methods. The case of brakeforming is best described as closed-loop property identification (of the M-K relationship) followed by open-loop control based on this identified model. For 3-D sheet deformation the necessary iteration is described as a discrete control system with a sample time of one forming cycle. The transfer function concept is then used to "identify" the forming characteristics of the workpiece which are then used in the next time step (forming cycle) of the control loop to change the die shape.

ROLL BENDING

This process will be treated first because it lends itself to the most conventional application of control theory. From Eqn (1) it can be seen that if
we measure K_{max} and M_{max}, using methods illustrated above, we can continuously calculate K_u, which is the desired process output. If we simply
construct a "curvature servo" as illustrated in Fig. 7, then it will be possible to continuously command and regulate the final, unloaded shape of the
part. Since the springback of the material is continuously measured, there
is no dependence upon prior knowledge of this property, and the material insensitive nature of this closed-loop process has been verified in experiments
on computer controlled prototype machines (2,3,6). Furthermore, it is apparent that this is an absolute curvature controller, and if the workpiece
fed to this process has a non-zero initial curvature (which manifests itself
as an offset of the origin of the M-K relationship), this "disturbance" will
be treated in the same manner as springback changes, and will rejected. As
pointed out by Hardt and Hale (7) this property can be exploited to allow
continuous, single pass straightening of aluminum extrusions, given only the
cross-section geometry of the workpiece.

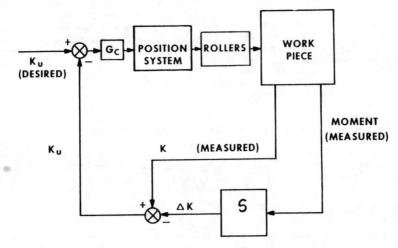

Figure 7   A Curvature Servo for Roll Bending

BRAKEFORMING

In the case of brakeforming, the desired output, final unloaded angle θ_u cannot be directly measured in a single pass strategy, so a closed-loop system
as shown in Fig. 7 is not feasible. Several approaches can be taken here,
but the one that has been demonstrated on a full size machine is the one described by Stelson (5). In this algorithm, the details of which can be found
in (5), the punch force F_p and displacement Y_p were measured continuously,
and used to construct an equivalent moment-curvature diagram. This was done

251

by assuming that the workpiece material behaved as an elastic-perfectly plastic material. This then leads to an analytical expression for the M-K relationship (for a wide, thin sheet) of:

$$M = K \, EI/(1-\nu^2) \qquad \text{for } M < M_y$$
$$M = 1.5 \, M_y \, (1- 1/3(K_y/K)^2) \qquad \text{for } M > M_y \tag{4}$$

Where I is the area moment of inertia, ν is Poisson's ratio, and M_y and K_y are the yield moment and curvature respectively.

By noting that this expression reaches an asymptotic value of $M = 1.5 \, M_y$ as $K \gg K_y$, and realizing that measuring F_p allows us to estimate M, we can estimate M_y by assuming that at large Y_p, the curvature is well beyond K_y. (This is also gauged by noting where yield occurs on the basic $F_p - Y_p$ data). Given M_y, and a measured or calculated value for EI, the entire expression in Eqn (4) is calibrated.

Once the M-K relationship is known, it is then assumed that the moment varies linearly from a (measured) maximum at the center of the forming region to zero at the die contact points. Equation (4) is then used to calculate the curvature distribution from the moment distribution, and the total angle in the section is calculated from Eqn (3). This then elucidates the entire loaded state of the sheet, and the springback is can be calculated by assuming that each point unloads elastically to zero moment. Thus, the actual control is exerted in a single stroke of the forming press, with the initial stages devoted to measuring F_p and Y_p, calculating the equivalent M-K relationship, and the resulting input-output relationship $\theta_u(Y_p)$. The process is then continued until the target value of θ_u is reached.

It can be seen that this approach involves an in-process material property identification followed by an *open-loop* control of the press based on that identification. As pointed out by Stelson, the major shortcoming of this approach is in describing the actual sheet curvature in the die region, which leads to errors in the $\theta(Y_p)$ estimate, but does not severly affect springback estimates. This implies (and it has been shown), that if the angle θ is measured directly, a highly accurate process can result.

THREE DIMENSIONAL FORMING

Control of the 3-D forming process is based on measurement of the sheet shape and working in the spatial frequency domain. As shown earlier, with shape information we can construct the deformation transfer function H according to Eqn (2). If on a given cycle of forming, (denoted by a subscript i) we know H_i, then the appropriate shape for the die for the next (i+1) cycle can be found from

$$D_{i+1} = D_i + H^{-1} \, (P_i - P_o) \tag{5}$$

where P_o is the desired or target part (described by its equivalent spatial frequency spectrum) and the term $(P_i - P_o)$ represents the part error after then ith cycle. (Note the incremental nature of this controller as implied by the presence of D_i on the right hand side of Eqn(5).

While this again is a control method that first identifies the forming characteristics of the workpiece (in this case via H_i) the fact that no force or stress information is available precludes single cycle forming, and an

252

iterative procedure is necessary. However, this iteration is augmented by a continuously updated estimate for H, which in turn continuously improves the die design for the next cycle. This is in fact a closed-loop strategy, and an equivalent control diagram is shown in Fig. 8, where the z^{-1} term represents a time *delay* of one forming cycle. The identification procedure is then shown to operate so as to continually vary the forward path of the controller so that the part error is used in the best way to modify D, the die shape.

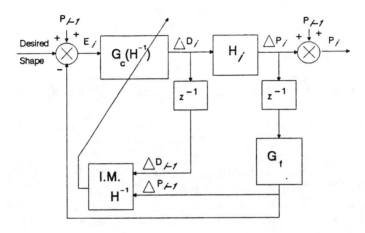

Figure 8   A Closed-Loop 3-D Forming System based on a Deformation Transfer Function (H_i) Identification, with a Single Cycle Delay (z^{-1})

This procedure has been applied to sheet stamping of axisymmetric parts, by first designing a system for rapidly re-machining die on a CNC lathe, forming parts with well controlled edge conditions, and then measuring the result on a coordinate measuring machine. The control loop for this system was then closed using a form of Eqn (5) and rapid convergence to the desired shapes has been demonstrated (8). Extension to fully general three dimensional shapes is being explored through development of a flexible tooling that can be reconfigured rapidly under computer control (9).

CONCLUSIONS

Sheet forming processes have many economic and engineering advantages over other processes, but are plagued by inaccuracy and setup cost associated with material property and forming environment uncertainties. This paper has addressed a basic approach to this problem that is founded on control system theory. By combining causal models of the relevant deformation processes with available in-process measurements, it is possible to resolve much of the inherent uncertainties in forming, and to develop processes that are insensitive to material property variations, and that require little or no setup. This then not only decreases production time, it open the process to production environments of low to medium batch sizes where previously it has not been applicable. Furthermore, the lack of sensitivity to material properties and forming boundary condition carries with it the implication that costly material processing steps necessary to guarantee uniform properties may not be necessary.

While the approach has been successfully applied to several process in the laboratory, and has seen some in-plant trials at full scale, there remains considerable research to be done in all three areas: modelling, measurement and real-time control.   When the basic feedback strategy advocated in this paper is adopted, the benefits can only come if the system behavior is well understood, and the robustness for which it is implemented can be assured.

REFERENCES

1. Lee, C., and Hardt, D.E., "Closed-Loop Control of Sheet Stability During Stamping ", _Proc. 9th N. American Manuf. Res. Conf._, May, 1986.

2. Hardt, D.E., Roberts, M.A., and Stelson, K.A., "Closed-Loop Control of A Roll Bending Process" _Journal of Dynamic Systems, Measurement and Control_, 104, no. 4, 1982.

3. Hale, M.A., and Hardt, D.E., "Dynamics and Control of a Roll Bending Process", to appear in _IEEE Control Systems Magazine, 1987._

4. Webb, R.D., and Hardt, D.E., "Control of 3-D Sheet Deformation:Algorithm Development", _Sensing and Control of Manufacturing Processes and Robotic Systems_, ASME, Nov, 1984.

5. Stelson, K.A., "An Elastic-Plastic Model for In-Process Springback Control in Brakeforming", _Journal of Engineering for Industry_, 1984.

6. Lee M., and Stelson, K.A., "Adaptive Control of a Straightening Process", _Sensing and Control of Manufacturing Processes and Robotic Systems_, ASME, New York, 1984.

7. Hardt, D.E., and Hale, M.A, "Control of a Roll Straightening Process", _Annals of CIRP_, 1985.

8. Hardt, D. E., and Webb, R.D. "Closed-Loop Control of Sheet Stamping: Algorithm Development Update" Laboratory for Manufacturing and Productivity Report no. FAR 86-5, 1986.

9. Hardt, D. E., Webb, R.D., and Robinson, R.E., "Closed-Loop Control of Sheet Forming: Algorithm Development and Machine Design", _Proc. 9th NSF Conference on Production Research_, May, 1985.

MANUFACTURING SCIENCE INTELLIGENT FABRICATION OF ICs IN THE 1990s

Dr. D. H. Phillips

Semiconductor Research Corporation[*]
4501 Alexander Drive, Suite 301
Post Office Box 12053
Research Triangle Park, NC 27709

Abstract

The mission of Manufacturing Sciences is the development of an understanding of semiconductor manufacturing processes to provide the quantification and control necessary to achieve a predictable and profitable product output in the competitive environment of the next decade. This mission includes not only technique research but our concern with the image of the manufacturing profession and the caliber of the graduate students who are attracted to the profession. Specifically, research goals are directed to quality, productivity, cost, IC packaging, CAD/CAM/CAT, and reliability. Manufacturing Sciences research includes efforts that address computer-aided manufacturing and advanced processing technologies, automation, yield enhancement and reliability, packaging, and metrology.

Manufacturing Sciences Principal Thrusts

The objective of the SRC Manufacturing Science program is to develop a university research base that enables available technical resources to create an optimum integrated circuit manufacturing capability at a competitive cost, quality, and reliability. Inherent in this objective is the development of scientific and engineering talent that can contribute to this field of knowledge and skills that can be applied in the semiconductor manufacturing industry. In furtherance of this objective, the principal thrust of the manufacturing sciences research programs that are in place as of September 1986 are given in Figure 1.

---

[*] The Semiconductor Research Corporation (SRC) is a cooperative, formed in 1982 by U.S. companies, that addresses generic research and educational needs (primarily through a university-based, contract research program) related to the design, development, and production of integrated circuits in order to enhance the competitiveness of its members.

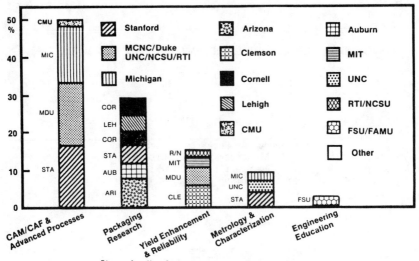

Figure 1. Manufacturing Sciences Principal Thrusts

In the research agenda of the SRC, manufacturing productivity and quality have been given high priority. Historically, little effort has been invested by U.S. universities in semiconductor manufacturing research. For this SRC initiative, the research scope has been defined, the attention of competent investigators has been obtained, and a research program is underway. Manufacturing Sciences research must progress at a rapid pace in the U.S. Neither the manufacturing tools nor the production know-how required to produce competitive mega-device chips is available. Advances are required in simulation of the equipments and processes, in the control of the integrated fabrication sequence, in automation, in testing, and in providing strong interfaces between the manufacturing and design environments. Manufacturing Science research is being carried out at Stanford (CIM/CAF), the Microelectronics Center of North Carolina (CMOS fabrication research), and the University of Michigan (expert systems, sensors, and machine vision). Results to date include demonstration of in-process sensors, including a thermal imager for temperature profiling and an end-point detector for plasma etching. Research is underway to develop a manufacturing simulation environment that includes models for processes (reactive ion etching), equipment, and material flow.

New Directions and Summary

A key issue that concerns a number of SRC members relates to the competitive threat to the U.S. semiconductor manufacturing equipment industry. To address this issue, a panel was organized under the aegis of the SRC Manufacturing Sciences Committee. In a series of meetings, problems are discussed and several innovative solutions have been proposed and are currently being implemented. One example is the evolution of a protocol for improving interactions between the builders and users of the equipment to promote better semiconductor-process-equipment products.

NOW   1990s

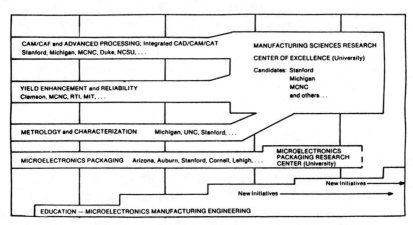

Figure 2. SRC Manufacturing Sciences Roadmap and Strategy

The new directions for the SRC Manufacturing Sciences research program are illustrated in Figure 2. The four principal thrusts (described above) are planned to continue, concentrating on a coalescence of research tasks to form a stronger benefit for the U.S. semiconductor industry.

Reference

1. D. H. Phillips, "Future Directions in Manufacturing Science: University/Industry Research and Education Needs of the American Semiconductor Industry in the 1990s," presented at the IEMT Conference, Session M-2B, San Francisco, CA, September 15, 1986.

Subject Index

260

Author Index